MW00760543

# Theory of Quantitative Magnetic Resonance Imaging

# Theory of Quantitative Magnetic Resonance Imaging

## Hernán Jara
Boston University, USA

 **World Scientific**

NEW JERSEY · LONDON · SINGAPORE · BEIJING · SHANGHAI · HONG KONG · TAIPEI · CHENNAI

*Published by*

World Scientific Publishing Co. Pte. Ltd.

5 Toh Tuck Link, Singapore 596224

*USA office:* 27 Warren Street, Suite 401-402, Hackensack, NJ 07601

*UK office:* 57 Shelton Street, Covent Garden, London WC2H 9HE

**British Library Cataloguing-in-Publication Data**
A catalogue record for this book is available from the British Library.

ISBN 978-981-4295-23-9

Typeset by Stallion Press
Email: enquiries@stallionpress.com

Printed in Singapore by B & Jo Enterprise Pte Ltd

*Para Sebastián y Carola,*
*mis compañeros de vida*

# CONTENTS

# PREFACE

*Quantitative Magnetic Resonance Imaging (qMRI)* refers to a diverse collection of image acquisition techniques and image processing techniques that are used as matched pairs for mapping the spatial distributions of the physical quantities that influence MRI signals; such physical quantities are known as qMRI parameters. The many qMRI parameters include: 1) measures of the amount of the MR-active substance --e.g. the proton density--. 2) Measures of states of motion --i.e. kinetic properties such as molecular diffusion, perfusion, and flow--, and 3) measures of interactions between the MR-active substance and the molecular environment --e.g. relaxation times and magnetization exchange parameter-- as well as interactions between the patient and the MRI scanner.

Succinctly, qMRI is the science of mapping -- or imaging -- qMRI parameters, or equivalently, the science of quantifying tissue properties at the spatial scale of imaging volume element (voxel) as represented by the picture element (pixel). Accordingly, the central mathematical object of qMRI is the numerical value of every pixel --or pixel value-- and the main objective of qMRI is to generate scientific grade pixel values that bear scientific units of measurement and that therefore have a more absolute meaning than the pixel values of directly acquired MR images. Most MR images currently used in clinical practice consist of non-quantitative pixel values meaning that these do not bear scientific units and are not directly comparable to pixel values of other images of even the same patient. Such directly acquired images are weighted by qMRI parameters and the pixel values have meaning only in relation to other pixel values in that particular dataset.

qMRI is an evolving scientific discipline that has the potential of impacting all stages of clinical and research MRI practices, from image acquisition, to image processing, to image interpretation. To the best of my knowledge, to date there are only two other books in this subject matter, specifically "Quantitative MRI of the Brain: Measuring Changes Caused by Disease", Edited by Paul Tofts and "Quantitative MRI in Cancer", Edited by William Hendee. These landmark books cover qMRI theory in a manner that is intertwined with medical applications. The purpose of the much shorter book herein is to provide a concise and unified theoretical description of qMRI theory only and is intended primarily as a textbook for a graduate level course potentially offered by academic departments such as Bioimaging, Biomedical Engineering, Computer Sciences, Mathematics, or Physics.

This textbook has its origin in lecture notes for an undergraduate course --Introduction to Medical Imaging-- that I have been teaching for the past eleven years as part of the curriculum of the Biomedical Engineering department at Boston University. It is through interactions with the students and with my colleagues of the Radiology Department at Boston University that I have come to understand the unifying principles of Medical Imaging, its implications to qMRI theory, and the value of quantification in medicine. I owe much to my research partners and colleagues Stephan Anderson, Joseph Ferrucci, Alexander Norbash, Naoko Saito, Osamu Sakai, Jorge Soto, and Memi Watanabe with whom I have collaborated for many years. I am also deeply grateful to Peter Joseph and Felix Wehrli who introduced me, in my formative years at the University of Pennsylvania, to the fields of MRI and qMRI. Finally, I wish to thank Stephan Anderson for editing this manuscript.

My dear father Alvaro Jara dedicated his academic life to research in the field of Quantitative History. He inculcated in me a deep appreciation for the power of quantification as a tool for understanding humankind as well as nature in its various dimensions. When he passed away some years ago, my dear friend and mentor Dr. Joseph Ferrucci told me that lost loved ones "reverberate" in time, and these kind words certainly reverberate in my mind as I finish this book.

HJ, 2013, Belmont, MA

# A. INTRODUCTION

## A1. Historical Notes

The earliest known manmade images are paintings in cave walls representing various animal and human figures. Although precise dating of these primitive images is not without uncertainty, the earliest paintings are believed to date to the prehistoric times, some as old as 32,000 years ago. Considering that our earliest human ancestors date back several million years, this is very recent in the human evolutionary timeline, thus indicating that the practice of imaging as a human activity is a manifestation of very advanced intellectual functions.

The camera obscura, originally described around 1000-AD, is the first human invention that could generate images artificially in a manner bypassing the human brain. The principle of operation of the camera obscura is very similar to that of the eye consisting of a sealed box with a pinhole via which specific rays of light are selectively accepted into the box. With this setup, only the rays of light making a point-to-point correspondence between the object and the imaging plane are accepted into the camera obscura thus forming a high-fidelity image of an object positioned in front of the pinhole. It would take another 800 years until the invention of the photographic camera; the first imaging device that could not only create images artificially but that could also store those permanently using plates coated with a light-sensitive silver-halide emulsion.

It would take another hundred years until the discovery of x-rays by Röntgen (Roentgen, 1896), the first form of radiation that had the prodigious property of penetrating solid matter in general and human tissues in particular, thus permitting for the first time the investigation of the internal human body by nondestructive means. So began a new branch of medicine, namely radiology, which initially was limited to producing shadow projectional images of the human body.

During the classical period of medical imaging --from 1896 until the early 1970's--, imaging scientists and pioneering imaging physicians investigated other penetrating radiations, contrast agents, and specialized scanning techniques and technologies thus widening the range of applications of this emerging medical imaging science. This classical period saw the advent of imaging with γ-rays emitted by radioactive nuclei (Cassen, Curtis, Reed, & Libby, 1951; Sweet, 1951; Wrenn Jr, Good, & Handler, 1951) and with ultrasonic radiation (Dussik, 1942; Woo, 2010); nuclear imaging and ultrasound imaging respectively.

In parallel to these scientific developments, and without an apparent connection to medical imaging, the field of nuclear magnetic resonance (NMR) emerged (Bloch, Hansen, & Packard, 1946; Purcell, 1946) and evolved into one of the most fruitful branches of physics and chemistry. The use of strong magnetic fields in conjunction with long-wavelength electromagnetic radiation (radiowaves) permitted nondestructively probing condensed matter *via* magnetic interactions with atomic nuclei, specifically with the magnetic moments of some atomic nuclei, and most notably with the $^1$H-proton nucleus of the hydrogen atom the most abundant chemical element in nature and the primary building block of biologic matter.

The contemporary period of medical imaging begins with the invention in 1972 of computed tomography (CT) with γ-rays (Chesler, 1973) and x-rays (Cormack, 1963; Hounsfield, 1973). In the case of x-ray CT, the new imaging device combined a movable-x-ray-beam transmission apparatus that targeted a single thin axial slice, with a digital computer. The CT scanner would generate many geometrical projections of the targeted slice at different angles of tomographic projection, which were stored sequentially in the permanent memory of a computer thus generating a full data set in signal space and allowing further mathematical processing. A reconstruction program based on the mathematics of continuous geometric projections --developed by Johann Radon in 1917 (Radon, 1917) -- would then transform the signal space data set into a two dimensional representation of the axial slice in geometric or anatomic space. The ingenious strategy of probing internally an object with successive and systematically different radiation experiments for generating a complete representation of a thin slice in signal space --and to later transform it into geometric space-- was soon thereafter adopted for tomographic imaging with the nuclear induction experiment (Lauterbur, 1973). In this case, spectral projections in Fourier domain were obtained by reading time dependent NMR signals while applying a magnetic field gradient; thus marking the

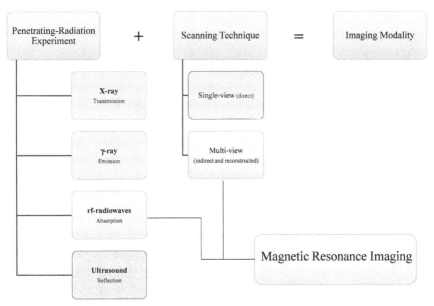

Fig. A-1. Classification of medical imaging modalities in terms of a penetrating-radiation experiment and a scanning technique: MRI results from combining the Nuclear Induction experiment with a multi-view scanning technique. The penetrating-radiation experiment uses electromagnetic radiation in the radiofrequency part of the electromagnetic spectrum that is absorbed by magnetically polarized matter. The induced signals are encoded spatially in multiple experiments and the resulting multiple views of the object are reconstructed to form an image.

beginning of Magnetic Resonance Imaging (MRI). Imaging modalities that require an intermediate reconstruction step for image formation will be referred to as indirect imaging modalities to distinguish them from the simpler direct imaging modalities of the classical period.

Altogether, by the early 1980's, a wide array of imaging devices based on each of the five primary penetrating radiation experiments and the two main scanning modes (projectional *vs.* computed tomography) become commercially available and gave shape and structure to the contemporary radiology department.

## A2. Image Processing

Using computers for image acquisition has proven invaluable for generating images of remarkable resemblance to the patient's actual internal anatomy; in particular, the high imaging fidelities afforded by contemporary x-ray

CT and MRI rivals in many cases the realism of photographs of *ex-vivo* frozen sections. However, the use of computers has not been limited to the image acquisition-reconstruction front end of medical imaging and these uniquely adaptable machines find numerous post-acquisition uses. The field of medical Image Processing (IP) is becoming a science in its own right with well-defined principles and theory. Of the many uses of IP, probably the best well-known methods are for improving image quality with respect to the four objective measures of image quality, *i.e.* signal-to-noise ratio (SNR), spatial resolution, contrast-to-noise ratio, and artifact reduction. Other well-known IP methods include methods for quantifying the sizes and shapes of organs, which include segmentation and volumetry, and methods for co-registering spatially different images from different scans.

## A3. Quantitative Imaging

Another branch of image processing known as Quantitative Imaging (QI) is gradually gaining acceptance in the scientific community (Tofts, 2003). The main objective of Quantitative Imaging is to elevate each pixel value of an image to a scientific-grade measurement that can be interpreted independently of other pixels and standardized across imaging scanner platforms, across imaging sessions, and ultimately standardized among different patients. The QI concept entails processing images on a pixel-by-pixel basis to remove the experimental information that is unrelated to the patient and is therefore superfluous, and to retain the pure physical content of the measurement as it relates to the tissue(s), which a pixel represents. Such superfluous information includes the scanner settings and the detection sensitivity factors. Accordingly, QI processing amounts to purifying the information of each pixel value down to the physical property probed with the radiation experiment of the modality. Images of all modalities can be QI-processed; for example by properly calibrating and intensity correcting the pixel values of an x-ray CT image, a map of the linear attenuation coefficient results that is universally expressed in Hounsfield units, which are related to the local density of electrons. Similarly, gamma-ray images can be QI-processed leading to maps of the radionuclide concentration or uptake.

Quantitative MRI (qMRI) is unique for several reasons: 1) it is multispectral affording many possible qMRI parameters per pixel. 2) These qMRI parameters characterize the states of water and lipids in tissue, and are therefore the most biologically-meaningful parameters that can be

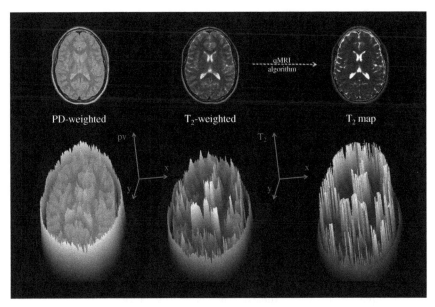

Fig. A-2. Pixel values with MRI *versus* qMRI: illustrates the concept of pixel value for two directly acquired images and for a qMRI $T_2$ map. In the first case, the pixel values are relative numbers devoid of scientific units and in the second, it is the actual $T_2$ relaxation time in units of seconds. The directly acquired images (left and middle) are used for computing the $T_2$ map (right). Images (top row) are also displayed as surface plots (bottom row) with pixel values indicated by arrows.

determined with any currently established medical imaging modality, and 3) all qMRI parameters other than the proton density itself, can be mapped using qMRI algorithms based on the principle of differential weighting. This principle, which stems directly from MRI physics, establishes a logical and powerful mathematical paradigm for removing the patient-unrelated experimental contributions of the pixel values.

## A4. Book Organization

This book is concerned with the underlying mathematical and physical principles and theories needed for understanding and practicing qMRI, including general imaging theory, the theory of NMR dynamics –Bloch equations and generalized Bloch equations—, spatial localization with magnetic field gradients, and the microscopic theory of NMR relaxation. This book is also concerned with practical issues related to the design of qMRI pulse sequences and the complementary qMRI algorithms, which should work synergistically for optimum qMRI accuracy and image quality. The last chapter of this book deals with insights into possible clinical applications of this emergent field.

## References

Bloch, F., Hansen, W., & Packard, M. (1946). Nobel Lecture: Nuclear induction. *Physical Review, 70*(7–8), 460–474.

Cassen, B., Curtis, L., Reed, C., & Libby, R. (1951). Instrumentation for I-131 use in medical studies. *Nucleonics, 9*(2).

Chesler, D. (1973). *Positron Tomography and Three-Dimensional Reconstruction.*

Cormack, A. (1963). Representation of a function by its line integrals with radiological applications. *J. Appl. Phys, 34*, 2722–2732.

Dussik, K. (1942). Ueber die Möglichkeit, hochfrequente mechanische Schwingungen als diagnostisches Hilfsmittel zu verwerten. *Zeitschrift für die gesamte Neurologie und Psychiatrie, 174*(1), 153–168.

Hounsfield, G. (1973). Computerized transverse axial scanning (tomography). 1. Description of system. *Br J Radiol, 46*(552), 1016–1022.

Lauterbur, P. C. (1973). Image Formation by Induced Local Interactions: Examples Employing Nuclear Magnetic Resonance. *Nature, 242*(5394), 190–191.

Purcell, E. (1946). Nobel Lecture: Research in nuclear magnetism.

Radon, J. (1917). On determination of functions by their integral values along certain multiplicities. *Ber. der Sachische Akademie der Wissenschaften Leipzig,* (Germany), *69*, 262–277.

Roentgen, W. (1896). Sitzungsberichte Würzburger Physik-medic. Gesellschaft, 137, 132–141, 1895; translation by Arthur Stanton as On a New Kind of Rays. *Nature, 53*, 274–276.

Sweet, W. (1951). *The uses of nuclear disintegration in the diagnosis and treatment of brain tumor*: AECU-1651, Massachusetts General Hospital; Brookhaven National Lab.

Tofts, P. (2003). *Quantitative MRI of the brain: measuring changes caused by disease*: John Wiley & Sons Inc.

Woo, J. (2010). A short history of the development of ultrasound in obstetrics and gynecology: http://www.ob-ultrasound.net/history1.html

Wrenn Jr, F., Good, M., & Handler, P. (1951). The use of positron-emitting radioisotopes for the localization of brain tumors. *Science, 113*(2940), 525.

# B. ELEMENTS OF IMAGING THEORY

## B1. Introduction

This chapter is concerned with the general theory of internal imaging with penetrating radiations as applicable to all medical imaging modalities, from spatial encoding, to spatial localization and image formation, to objective measures of image quality. The goal is to present formulae that later will be directly applicable to spatial localization with MRI.

## B2. Imaging as a Mathematical Operation

### B2.1. *Spatial Encoding*

An imaging experiment is a sequence of laboratory procedures by which a material object is probed internally with penetrating radiations. The purpose is to generate measurements of spatially encoded signals that can be subsequently back-transformed or equivalently reconstructed into geometric space. In this way, spatially resolved representations are generated that portray the internal spatial distribution of one of the object's physical properties. Examples of such physical properties include the linear attenuation coefficient in x-ray CT, the uptake concentration of a radiopharmaceutical in $\gamma$-ray CT, and the transverse magnetization in MRI.

Nondestructive internal imaging requires generating spatially encoded signals that build up a different mathematical space termed herein signal space. The goal is to design image acquisition techniques such that the signal space is conjugate to geometric space in a strict mathematical sense. Ideally, the mathematic transform should be easily computable – *e.g.* Fourier and Radon transforms-- so that the reconstruction algorithms can efficiently reorganize the signal space data back into geometric space. The tool that performs the reorganization of the signal data back to geometric space data is the so-called reconstruction program. We will use the symbol $(\vec{K})$

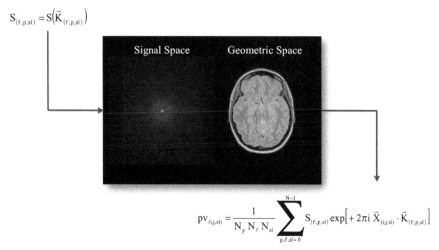

$$S_{(f,p,sl)} = S\left(\vec{K}_{(f,p,sl)}\right)$$

$$pv_{(i,j,sl)} = \frac{1}{N_p\,N_f\,N_{sl}} \sum_{p,f,sl=0}^{N-1} S_{(f,p,sl)}\,\exp\left[+2\pi i\,\vec{X}_{(i,j,sl)}\cdot\vec{K}_{(f,p,sl)}\right]$$

Fig. B-1. Signal space data and geometric space data in the case of Fourier transform MRI are shown together with the corresponding defining equations. In this example, the signal space is the Fourier domain representation of the object.

for the position vector of data points in signal space with the understanding that the $(\vec{K}, \vec{x})$ pairs formed together with the position vector in geometric space are conjugate variables in a strict mathematical sense according to the mathematical transformation linking the signal space to the geometric space.

Accordingly, internal imaging experiments can be modeled mathematically as continuous-to-discrete operations that transform one of the object's internal fields — generically represented throughout this book by a continuous function of space $h(\vec{x};\tau)$– into a discrete array of numbers, referred to as pixel values (pv).

$$h(\vec{x};\tau) \xrightarrow[\text{Experiment}]{\text{Image}} S(\vec{\xi}_{(i,j,k)}) \xrightarrow[\text{Reconstruction}]{\text{Image}} pv(\vec{X}_{(i,j,k)})$$

$$= \Omega_{(i,j,k)} \iiint_{\left\{\Delta V_{(i,j,k)}\right\}} h(\vec{x};\tau)\,d^3x \qquad \text{(Eq. B2.1–1)}$$

Accordingly, the resulting pixel values are proportional to the integral of the object's internal field over small sub-volumes known as imaging volume elements (voxels). Furthermore, in this model of an internal-imaging experiment, the symbols $pv_{(i,j,k)}$, $\Omega_{(i,j,k)}$, and $\Delta V_{(i,j,k)}$ are respectively the pixel value, the detection sensitivity function and the voxel volume at the $(i,j,k)$-grid position. Moreover, $\tau$ represents the time variable in cases in which the physical property studied depends explicitly on time, such is the case of the transverse magnetization in MRI.

The resulting voxels and their associated pixel values are arranged in geometric space according to an imaging grid, as defined by an array of position vectors such as:

$$\vec{X}_{(i,j,sl)} = i\Delta x \, \hat{x} + j \, \Delta y \, \hat{y} + k \, \Delta z \, \hat{z} \qquad \text{(Eq. B2.1–2)}$$

where $(\hat{x}, \hat{y}, \hat{z})$ is a set of unitary vectors defining the three orthogonal directions of geometric space, $(\Delta x, \Delta y, \Delta z)$ are the pixel spacing along each direction, and the integer indices run from one to the respective matrix sizes $(N_x, N_y, N_z)$ in each direction respectively. We note that for the direct imaging modalities of the classical period, the spatially encoded signal measurements are the same as the pixel values thus the intermediate reconstruction step is unnecessary.

## B2.2. *Spatial Localization: the Voxel Sensitivity Function*

To a first approximation, voxels can be modeled as cuboids of linear dimensions equal to the pixel spacing of the imaging grid along each direction. Although the cuboid voxel interpretation is useful for many image analysis purposes such as signal-to-noise and pulse sequence optimization calculations, it is an idealization and therefore not strictly valid. In particular, the imaging effects of sub-voxel tissue structure and sub-voxel motion are not faithfully described with the cuboid voxel model. A more rigorous mathematical formalism based on the concept of a voxel sensitivity function (VSF) was formulated by Parker et al. (Parker, Du, & Davis, 1995) to study the sensitivity of conventional Fourier transform (FT) MR imaging pulse sequences to tissue structures with subvoxel dimensions.

The VSF model leads to a useful alternative form of the pixel value equation above (see Eq. B2.1–1), which represents more intuitively the spatial localization process. In this alternative form, pixel values can be expressed as a spatial integral over infinite space, and the process of spatial localization is accounted for automatically by the VSF that acts as a spatial filter. This spatial filter nullifies the signal contributions to the pixel value that would arise from tissues outside the voxel. Accordingly, we can write:

$$pv_{(i,j,k)} = \Omega_{(i,j,k)} \iiint\limits_{\left\{\substack{\text{Infinite}\\ \text{space}}\right\}} h(\vec{x}; \tau) \, VSF(\vec{X}_{(i,j,k)} - \vec{x}) d^3x \qquad \text{(Eq. B2.2–1)}$$

The problem then becomes one of finding the precise VSF as dictated by the scanning technique's spatial encoding method. As we will study later, in the case of MRI and under typical imaging conditions, the VSF shape can be calculated in closed form (see Fig. B-2).

Furthermore, by recognizing that the triple integral represents a convolution operation in three dimensions, the net effect of an imaging experiment can be modeled as follows.

$$\underbrace{h(\vec{x}; \tau)}$$
$$\updownarrow$$
$$\{\vec{x} \in \text{Imaging Volume}\}$$

$$\xrightarrow{\text{i,j,k} \in (1..N_{x,y,z})} \quad \underbrace{pv_{(i,j,k)} = \Omega_{(i,j,k)} [h(\vec{x}; \tau) * * * \text{VSF}(\vec{x})](\vec{X}_{(i,j,k)})}$$
$$\updownarrow$$
$$\{\vec{X}_{(i,j,k)} \in \text{Imaging Grid}\}$$

(Eq. B2.2–2)

Hence, from a purely mathematical standpoint, imaging systems may be thought of as devices that perform a convolution operation between the object fields with a localized voxel sensitivity function and scan sequentially or in parallel along the discrete grid-positions inside the object. Accordingly, medical imaging systems can be conceptualized as convolving-scanning devices.

### B2.3. *Fourier Transform Imaging*

In MRI, the connection between spatial encoding with magnetic field gradients and Fourier transform mathematics stems from solving the equations of motion that describe nuclear magnetization dynamics in tissue. These are differential equations known as the Bloch equations and will be studied in section C3-5.

More specifically, as will be shown in sections C5.2 and C5.3, the Bloch equation solution for the complex transverse magnetization in the presence of magnetic field gradients is of the general form:

$$m(\vec{x}, te_{(f,p,sl)}) = M_z^{(eq)}(\vec{x}) \exp\left(-\frac{te_{(f,p,sl)}}{T_2(\vec{x})}\right) \exp\left(2\pi i \, \vec{x} \cdot \vec{K}_{(f,p,sl)}\right)$$

(Eq. B2.3–1)

$$VSF^{(2D)}(x,y) \propto \frac{1}{N_x N_y} \frac{\sin\left(\frac{\pi(X_{(i,j,sl)}-x)}{\Delta x}\right)}{\sin\left(\frac{\pi(X_{(i,j,sl)}-x)}{N_x \Delta x}\right)} \frac{\sin\left(\frac{\pi(Y_{(i,j,sl)}-y)}{\Delta y}\right)}{\sin\left(\frac{\pi(Y_{(i,j,sl)}-y)}{N_y \Delta y}\right)}$$

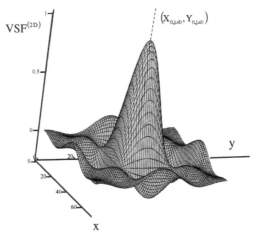

Fig. B-2. Voxel Sensitivity Function (VSF) for a 2D Fourier transform MRI pulse sequence. In the formula above, the VSF is expressed in terms of the matrix sizes in each direction ($N_x$, $N_y$), the coordinates of the position in the imaging grid (dotted line) and the pixel spacings in each direction. This formula will be studied in section C5.2.

where the quantities $M_z^{(eq)}(\vec{x})$ and $T_2(\vec{x})$ are the equilibrium longitudinal magnetization and the transverse relaxation time respectively. Most importantly for the present discussion, the vectors $\vec{K}_{(f,p,sl)}$ are functions of the applied magnetic field gradients and these are labeled with integer indexes that count the encoding steps along the so-called frequency encoding (f), phase encoding (p), and slice encoding (sl) directions (see section C5). This means that by applying magnetic field gradients in a judicious manner, the MR signals can be expressed as follows:

$$h(\vec{x},\tau) = m(\vec{x},\tau) \xrightarrow[\text{k-space}]{\text{Fourier Transform}} S(\vec{K}_{(f,p,sl)}) \propto \underset{\left\{\substack{\text{Infinite}\\\text{space}}\right\}}{\iiint} M_z^{(eq)}(\vec{x})$$

$$\times \exp\left(-\frac{te_{(f,p,sl)}}{T_2(\vec{x})}\right) \exp\left(2\pi i\, \vec{x} \cdot \vec{K}_{(f,p,sl)}\right) d^3x \qquad \text{(Eq. B2.3–2)}$$

This fundamental result establishes a Fourier conjugate relationship between geometric space and signal space. Furthermore, the signal equation above is the starting point for deriving pixel value equations as applicable to MRI pulse sequences, specifically:

$$pv_{(i,j,sl)} = \frac{1}{N_p\,N_f\,N_{sl}} \sum_{p,f,sl=0}^{N} S_{(f,p,sl)} \exp\left[+2\pi\,i\,\vec{X}_{(i,j,sl)} \cdot \vec{K}_{(f,p,sl)}\right]$$

(Eq. B2.3–3)

Accordingly, MRI pixel values are complex numbers, hence:

$$pv_{(i,j,sl)} = \mathrm{Re}\left\{pv_{(i,j,sl)}\right\} + i\,\mathrm{Im}\left\{pv_{(i,j,sl)}\right\}$$ 
(Eq. B2.3–4)

whereby the real and the imaginary parts stem from the same MR signal as components which are separated according to their phase relative to the master-clock signal reference of the MRI scanner. This is accomplished by means of specialized phase sensitive detectors.

Alternatively, pixel values can be expressed in terms of their magnitude and phase:

$$pv_{(i,j,sl)} = |pv_{(i,j,sl)}|\,\exp(i\,\Phi_{(i,j,sl)})$$
(Eq. B2.3–5)

The phase information is usually discarded in clinical practice. Most images used for diagnostic purposes are magnitude images and these are positive-valued.

Most MR imaging experiments are designed so that the resulting pixel values would be real numbers. In practice however, phase shifts develop for a variety of reasons (Henkelman, 1985) including scanner imperfections --eddy currents stemming from the imaging gradient fields-- and nonlinearities caused by interactions between the patient and the magnetic fields of the MRI scanner. Moreover, for some qMRI applications (see section E3), phase shifts are generated intentionally for quantifying certain qMRI parameters *via* the phase of the transverse magnetization. Such phase shift based qMRI techniques are useful for quantifying flow, spatial displacement, temperature and any other parameter that can cause a phase shift, as will be studied in chapter E.

## B2.4. *Noise*

The noise affecting each detection channel of an MRI system is generally assumed to derive from a bivariate normal distribution with a mean value of zero and a standard deviation $\sigma_g$. In other words, the noise of each detection channel has a gaussian probability density function of the form:

$$P(n_{R,I}) = \frac{1}{\sqrt{2\pi\, \sigma_g^2}}\, \exp\left(-\frac{n_{R,I}^2}{2\,\sigma_g^2}\right) \qquad \text{(Eq. B2.4–1)}$$

where, the subscripts R, I are used to label the two detection channels, which correspond to the real and imaginary parts of the generated pixel values.

The pixel values including the contribution of noise can be expressed in terms of the ideal noiseless pixel values *via*:

$$\mathrm{pv}_{(i,j,sl)} = \mathrm{pv}_{(i,j,sl)}^{(0)} + n_{(i,j,sl)} \qquad \text{(Eq. B2.4–2)}$$

where $n = n_R + i\, n_I$ and we have used the superscript $(0)$ to denote the noiseless pixel values. Furthermore, the noise contributions for the two in-quadrature channels $n_{R,I}$ are assumed uncorrelated to each other.

On the other hand, when considering magnitude images, the probability density function of noise is described by the Rician distribution (Gudbjartsson & Patz, 1995), which is given by:

$$P(m) = \frac{m}{\sigma_g^2}\, \exp\left(-\frac{(\mathrm{pv}^{(0)})^2 + m^2}{2\,\sigma_g^2}\right) I_0\left(\frac{\mathrm{pv}^{(0)}\, m}{\sigma_g^2}\right) H(m) \qquad \text{(Eq. B2.4–3)}$$

where the function:

$$I_0(z) = \frac{1}{2\pi} \int_0^{2\pi} \exp[z\, \cos(\alpha)]\, d\alpha \qquad \text{(Eq. B2.4–4)}$$

is the zero[th] order Bessel function of the first kind and $H(m)$ is the Heaviside step function, which ensures that the probability density function of a noisy magnitude MR image is zero for negative values of m.

## B3. Objective Measures of Image Quality

### B3.1. *Signal-to-Noise Ratio*

The signal-to-noise ratio (SNR) of a specific tissue or structure in an image is an objective measure of the sensitivity of the imaging technique for detecting signals stemming from that tissue, in relation to the random signal fluctuations as represented by image noise.

Experimentally, $SNR_{tissue}$ can be estimated from an image by means of region-of-interest (ROI) measurements, as the mean of the pixel values within an ROI positioned in the tissue of interest divided by the standard deviation of the noise:

$$SNR_{tissue} = \frac{\langle |pv_{(i,j,sl)}| \rangle_{ROI}}{stdev(noise)} \qquad \text{(Eq. B3.1–1)}$$

where the bracket signifies average over all pixels in the ROI. In turn, the standard deviation of the noise can be estimated for example with a second ROI positioned in an area devoid of signal and artifacts, such as the air around the patient. Visual perception of global image noise may vary from observer to observer depending on quality of vision and age. Nevertheless, as a rule of thumb, images appear noticeably noisy when the SNR of the brightest tissue falls below about a value of ten.

### B3.2. *Spatial (Geometric) Resolution*

The spatial resolution of an image is another objective quality measure concerning the ability of an image for depicting the finner details of the object. Alternatively, it is also a measure of the perceived sharpness with which all borders and delineation of tissue interfaces are portrayed. Spatial resolution is often referred to as high contrast resolution because this measure can be used to predict whether two small structures with significantly different pixel values relative to noise, can be resolved spatially. Conversely, if two adjacent tissues have nearly the same pixel values relative to the noise level, high spatial resolution alone is not sufficient to resolve them apart.

The original parameter used for quantifying spatial resolution along any given direction in space was derived from an instrument used for x-ray imaging equipment. This instrument consisted of a succession of increasingly thinner pairs of parallel bars. Each bar pair was made of two materials side by side, one transparent and one opaque to x-rays. As the bar pairs become increasingly thinner, these experiments resulted in lines in the test image and therefore spatial resolution was originally expressed in units of line pairs per millimeter. As such, spatial resolution is inversely related to the voxel dimension in that direction via:

$$R_{(x)}(lp/mm) = \frac{1}{2\Delta x} \qquad \text{(Eq. B3.2–1)}$$

where the factor of two in the denominator reflects the fact that a spatial resolution of one lp/mm requires at least one pixel in each of the 0.5 mm lines.

In theory and for a one dimensional object, two pixels are necessary to minimally resolve structure of length ($L_x$), in other words the voxel dimension need is

$$\Delta x \leq \frac{L_x}{2} \qquad \text{(Eq. B3.2--2)}$$

In practice however, operating in the regime of 5-10 voxels per structure is needed to resolve clearly a structure to counteract effects of blur, minor motion, and noise limitations.

Generalizing one dimensional spatial resolution considerations to three dimensions is greatly aided by considering not only the voxel volume but also its aspect ratio. For example, a voxel size typical of a lateral projectional x-ray of the head will be approximately equal to $0.1 \times 0.1 \times 100 = 1\,\text{mm}^3$, which is the same volume as one would obtain from a high resolution MRI with isotropic voxel at $1 \times 1 \times 1 = 1\,\text{mm}^3$. Even though the in-plane spatial resolution of such an x-ray image is tenfold higher relative to the MRI, the spatial resolution of the later is much superior because the in plane *vs.* slice thickness aspect ratio is 1,000 for the x-ray and 1 for the MRI.

$$\text{Aspectratio}_{(x)} = \frac{\Delta z}{\Delta x} \qquad \text{(Eq. B3.2--3)}$$

### B3.3. *Contrast-to-Noise Ratio: Pixel Value Resolution*

The third objective measure of image quality refers to pixel values differences or contrast between two tissues A and B; it is commonly expressed in relation to the noise level present throughout the image by the so-called contrast-to-noise ratio $\text{CNR}_{(A,B)}$, which is the difference of the individual signal-to-noise ratios, *i.e.*:

$$\text{CNR}_{(A,B)} = \text{SNR}_A - \text{SNR}_B \qquad \text{(Eq. B3.3--1)}$$

CNR is also referred to as low-contrast object detectability, which as the term implies gives a measure of detectability for objects with pixel values similar to that of its surroundings in the image.

## B3.4. *Vulnerability to Artifacts (MRI)*

Artifacts refer to unwanted image features with no correlation in the imaged object. These features, which arise from imperfections in the imaging technique and/or hardware limitations, may appear in an image as artificial intensity patterns, geometric distortions, or spurious ghosting. From a theoretic standpoint, artifacts result when the actual physical conditions at any time during the imaging experiment deviate from the ideal imaging conditions that constitute the reconstruction algorithm hypotheses. In practice, the most common artifacts encountered with modern MRI scanners --in proper working conditions-- originate from one of the following:

- Magnetic field inaccuracies (susceptibility artifacts).
- Spatial encoding errors due to tissue motion (ghosting artifacts).
- Existence of multiple resonant frequencies (chemical shift artifacts).
- Inadequate spatial resolution (truncation artifacts).
- Tissues extending beyond the imaging volume along the phase encoding (PE) direction (aliasing artifacts).

The typical image appearance of these artifacts, their causes, and the measures commonly used to alleviate their adverse effects, are reviewed in the following.

### B3.4.i. *Susceptibility artifacts*

These artifacts result from deviations in the value of the magnetic field inside the patient, during the application of the pulse sequence. Since modern scanners generate very homogeneous polarizing magnetic fields and highly linear magnetic field gradients, inaccuracies in the values of these are caused mainly by magnetic susceptibility variations within the patient. When present, these magnetic susceptibility variations distort the parent magnetic field as well as the magnetic field gradients used for spatial encoding. Because the function of the parent magnetic field is very different to that of the magnetic field gradients, the resulting artifacts may be categorized accordingly, *i.e.* first, inaccuracies of the main magnetic field cause a faster decay rate (*i.e.* shorter $T_2^*$) of the NMR signals in fast-field-echo pulse sequences. In regions of large magnetic field variations, the shortening of $T_2^*$ may lead to complete signal loss even at the shortest TE possible, as determined by the capabilities of the imaging subsystem. These so-called susceptibility artifacts are substantially alleviated with spin-echo pulse sequences because of the rephasing effect of the 180° refocusing pulse. Second, inaccuracies in the magnetic field gradients deteriorate the quality of the spatial encoding process thus potentially leading to a number of spatial encoding artifacts, including geometrical deformation, spatial misregistrations, and partial signal loss due to incomplete gradient balance. These artifacts are most noticeable when the susceptibility gradient magnitude is a significant fraction of the imaging gradient. Susceptibility gradients are strongest near the interface between two materials with different magnetic susceptibility. Susceptibility artifacts are therefore more prominent near anatomic interfaces separating two magnetically dissimilar materials. Important examples areas soft tissue-air interfaces near the nasal cavities, bone marrow-calcified bone, and soft tissue-metal in the case of patients with metallic orthopedic pieces.

### B3.4.ii. *Motion artifacts*

Movement of tissues during scanning interferes with the spatial encoding processes used in the pulse sequence. Because the time interval between successive measurements in the frequency encoding (FE)-direction is several orders of magnitude shorter than in the PE-direction, motion artifacts are almost exclusively observed along the phase encoding direction. Depending on the properties of flow and on the pulse sequence parameters, motion related artifacts might appear as image intensity smearing or as well-organized fainter replicas of the moving organs, which are referred to as ghosts. Vessel ghosting due to flowing blood is also a common occurrence. Sources of motion artifacts active during the time scale of conventional MRI include (from slowest to fastest): involuntary patient motion, respiration, peristalsis, vascular flow, and cardiac pulsations.

A number of methods have been used with varying degrees of success to counteract the adverse effects of motion in MRI. Some of these methods including increasing the number of signal averages (NSA), fat suppression, presaturation pulses, ordered phase encoding, and gradient moment nulling, may be referred to as indirect methods because they do not counteract directly the main cause of the problem which is the acquisition of imaging information with the organs in different positions at different times. Instead, indirect methods are used to reduce the severity of motion artifacts by either averaging (increasing NSA), eliminating the signals and consequently the ghosts of moving tissues which are not relevant for diagnosis (fat suppression and presaturation pulses), or by (partially) correcting the erroneous phases caused by motion (ordered phase encoding and gradient moment nulling). Alternatively, direct motion-artifact reduction techniques, which include respiratory triggering, cardiac triggering and breath holding, are designed so that the imaging data is generated with all tissues in the same location at all times.

### B3.4.iii. *Chemical shift artifacts*

The Larmor frequency of $^1$H-protons varies depending on the structure of the hydrogen-containing molecules. This phenomenon, which is known generically as chemical shift, is a direct consequence of the magnetic field-tracking property of the Larmor frequency. In the case of different chemical species, the local magnetic fields differ because the surrounding electron clouds modify differently the magnetic field at the site of the proton. In biological tissues, the two proton-containing chemical species of interest are water and lipids. The Larmor frequency difference between protons in water and protons in lipids is a very small fraction of the parent magnetic field, approximately 3.5 parts-per-million. In a 1.5 Tesla magnetic field, this translates to approximately a 220 Hz frequency difference. Therefore, the chemical shift artifact consists of an apparent spatial offset of the lipid-containing voxels relative to the water-containing voxels in the frequency encoding direction. This artifact is also referred to as chemical shift spatial offset.

### B3.4.iv. *Truncation artifacts*

These artifacts, which manifest in an image as ripples that occur near a sharp tissue interface in a parallel linear pattern, result from undersampling the tissue interface in the Fourier domain. Consequently, their severity decreases with decreasing voxel size.

### B3.4.v. *Aliasing artifacts*

An anatomical region of the patient that extends beyond a border of the field-of-view appears the opposite border, superimposed to the desired image. Aliasing artifacts are a consequence of the limitations of the phase encoding and the frequency encoding methods, whereby tissues extending beyond the borders of the field-of-view are tagged with integer multiples of the phases and of the frequencies associated with the tissues within the field-of-view, and are indistinguishable to the scanner. In practice, only aliasing artifacts in the phase encoding direction are of concern to the user because the extra frequency content in the MRI signals is eliminated electronically at a hardware level by means of filtering. Unfortunately, no such filtering means are available to eliminate selectively the extra phase content of the MRI signals, and consequently alternative measures are implemented to counteract the adverse effects of aliasing artifacts in the phase encoding direction. These include: 1) the foldover suppression method whereby the field-of-view is doubled in size during the scan and only the tissues contained in the desired field-of-view are displayed, 2) the use of receiver coils with sensitivity extending only slightly over the desired filed-of-view, and 3) spatial presaturation pulses, which reduce the MRI signals stemming from tissues beyond the desired field-of-view.

## References

Gudbjartsson, H., & Patz, S. (1995). The Rician distribution of noisy MRI data. *Magnetic Resonance in Medicine, 34*(6), 910–914.

Henkelman, R. (1985). Measurement of signal intensities in the presence of noise in MR images. *Medical physics, 12*(2), 232.

Parker, D. L., Du, Y. P., & Davis, W. L. (1995). The voxel sensitivity function in Fourier transform imaging: applications to magnetic resonance angiography. *Magn Reson Med, 33*(2), 156–162.

# C. PHYSICS OF QUANTITATIVE MRI

## C1. Introduction

This chapter is concerned with the theoretical physics platform of qMRI, from pure quantum mechanics for describing individual $^1$H-protons in water molecules, to the semi-classical description of the magnetization vector at the spin packet level *via* statistical mechanics, and finally to the voxel scale *via* imaging theory, as discussed in the previous chapter.

### C1.1. *The Different Spatial Scales of qMRI Theory*

Altogether, the theoretical framework of qMRI theory deals with spatial scales spanning about twelve orders of magnitude. Starting from $^1$H-protons at a spatial scale of $10^{-6}$ nm, to that of the water molecule at a spatial scale of $10^{-1}$ nm, to the somewhat undefined spatial scale of spin packets of $\sim$10 nm, and finally to the spatial scale of an imaging voxel, by current clinical standards of less than $10^6$ nm ($<$1 mm).

The main objective of this chapter is to lay out the equations of motion that are valid at each one of these spatial scale regimes, progressively from smallest, to intermediate, and finally to the imaging scale where qMRI image processing takes place. Particular effort is devoted at providing logical theoretical transitions from one spatial regime to the next by analyzing selected physical systems that constitute the building blocks for the next scale and by using mathematical notations that are intuitive and consistent among spatial scales. Particular attention is given to the study of the equations of motion for a *solitary $^1$H-proton in a time dependent magnetic field*, which are very similar to the Bloch equations and therefore are particularly useful for modeling the relaxation times in terms of microscopic randomly fluctuating magnetic fields, as discussed in Chapter D.

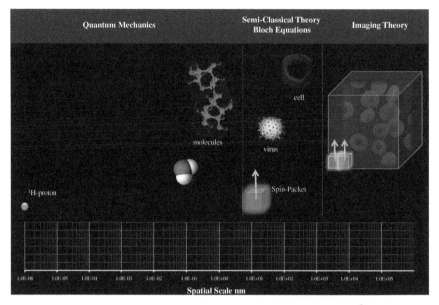

Fig. C-1. The physical theories and the spatial scales of MRI: From $^1$H-protons, to water molecules and macromolecules, to spin packets, and finally to imaging voxels.

## C1.2. *The Spatial Scale of the $^1H$ Proton and the Water Molecule*

Protons are stable spin-1/2 particles with a charge radius of approximately $0.875 \ 10^{-15}$m and possess a magnetic moment of 2.79 nuclear magnetons ($\mu_N = e\hbar/2m_p = 5.050 \ 10^{-27} \, \mathrm{JT}^{-1}$). This high magnetic moment as well as the high $^1$H-proton abundance in biological tissue, makes it uniquely useful for MRI in general and consequently for qMRI.

Quantum mechanics is necessary for describing $^1$H-protons, water molecules, and macromolecules. One can transition to semi-classical physics for the description of spin packets and voxels.

$$\underbrace{\delta V_{(1_{H-proton})} \longrightarrow \Delta V_{(H_2O)}}_{Quantum \ Physics} \xrightarrow{Statistical \ Mechanics} \underbrace{\delta V_{(packet)} \longrightarrow \Delta V_{(voxel)}}_{Semi-Classical \ Physics}$$

(Eq. C1.2–1)

The water's molecular geometry can be environment dependent; quoting Chaplin (Chaplin, 2010): "The experimental values for the gaseous water molecule are O–H length 0.95718Å, H–O–H angle 104.474°. These values are not equal in liquid water, where *ab initio* calculations (O–H length 0.991Å, H–O–H angle 105.5°) and neutron diffraction studies (O–H length 0.970Å, H–O–H angle 106°) suggest slightly greater values, which result from hydrogen bonding weakening the covalent bonding. These bond lengths and angles are likely to change, due to polarization shifts, in different hydrogen-bonded environments and when the water molecules are bound to solutes and ions."

We can calculate the intra-molecular proton-to-proton distance to be 1.55Å using the neutron diffraction values (O–H length 0.970Å, H–O–H angle 106°) and the triangle formula:

$$\beta = 2\,R_{OH}\sin\left(\frac{\phi}{2}\right) \qquad \text{(Eq. C1.2–2)}$$

The inter-molecular proton-to-proton closest distance of approach can be derived from published results of *ab initio* calculations as twice the longitudinal van der Waals radius of hydrogen in the water molecule, specifically 2.4Å. Although evidence exist that the water molecule is not perfectly spherical with about a ±5% variation in van der Waals diameter, here we assumed molecular sphericity and a van der Waals radius of 1.7Å. This value can be chosen for simulations because it is intermediate between the published minimum and maximum values of 1.6Å and 1.8Å respectively. We can assume to a first approximation that the shape and dimensions of water are the same in pure liquid water as in biological tissues. In other words, we can assume that water's electronic configuration does not change significantly due to interactions with the biological microenvironment.

## C1.3. *Spin Packets*

The spatial scale of spin packets is somewhat ambiguously defined; on one hand, the semi-classical NMR theory requires spin packets to be much larger than the water molecule so that number of contained spins is sufficiently large and the laws of statistical mechanics apply for describing the packet's spin state populations. On the other hand, the theory also calls for sub-microscopic spin packets so that the magnetization distribution inside the patient as well as inside the imaging voxels can be modeled as a continuous function of space and time, thus enabling theory formulation in terms of partial differential equations. In a crude model, we can imagine spin packets as

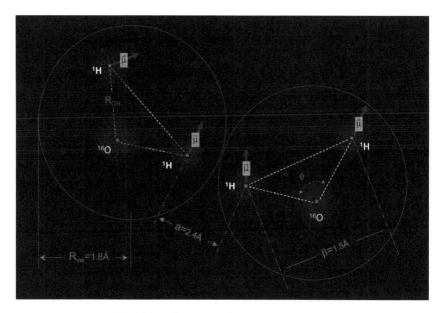

Fig. C-2.   Geometry of the water molecule.

small cubes containing about $10^4$ $^1$H-protons in order to fully guarantee the
appropriateness of the statistical mechanics description. Considering that
the water molecule diameter is approximately 0.34 nm, the linear dimension
of such spin packet can be estimated to be about 6-12 nm. Accordingly, by
defining an NMR theory spatial scale of ~10 nm for "continuous" NMR
description validity and noting that this is smaller than the size of the finest
sub-cellular structures (*e.g.* thinnest axons) as demonstrated with high-
resolution electron microscopy, we can safely assume that NMR spin packets
do not contain closed tissue compartments. We can therefore assume that
spin packets contain at most two types of water: free water and hydration
water. Furthermore, we can also assume that individual spin packets exhibit
essentially isotropic diffusion characteristics for times shorter than the
migrational time to adjacent spin packets. However, in the much longer
time scale of up to 100ms, which is the typical of the MRI echo times used
in clinical practice, a water molecule in pure water will self-diffuse by a
length:

$$L = \sqrt{6Dt}$$
(Eq. C1.3–1)

which approximates $36\,\mu$m using a value of $2.2 \ 10^{-3}\,\text{mm}^2/\text{s}$ for water at room temperature. Hence, during the typical measurement time of an MRI scan, individual water molecules can migrate on average about three thousand and two thousand spin packets in pure water and tissue, respectively.

The spatial scale of imaging voxels ranges from about $5\,\mu$m at the highest spatial resolution with highly specialized research scanners to the contemporary typical clinical voxel scale of less than $1\,\text{mm}$.

### C1.4. *The Magnetic Fields of MRI*

Besides the main magnetic field $B_0$ which is static and is used for longitudinally polarizing the patient tissues, applying additional magnetic fields is necessary in order to generate MRI signals, which are spatially encoded. The two types of time dependent magnetic fields used in MRI pulse sequences are:

1) The field used for manipulating the magnetization, which is known as the active $B_1$ field. We will later study the precise characteristics of such fields for the specific purposes of transverse magnetization generation (excitation), rf refocusing for generating the so-called spin-echoes, and also for achieving longitudinal magnetization inversion. For the time being, we describe generically the active $B_1$ fields as pulse-modulated, approximately harmonic, and circularly polarized in the transverse plane for efficient interaction with the magnetization, in other words:

$$\vec{B}_1^{(\text{rf})}(t) = B_1(t)(\cos(\omega_{\text{rf}}t)\hat{x} + \sin(\omega_{\text{rf}}t)\hat{y}) \quad \text{(Eq. C1.4-1)}$$

We shall see later that these oscillatory fields have the distinct property of rotating the magnetization vector trough the phenomenon of nuclear magnetic resonance when the carrier frequency is chosen equal to the *Larmor frequency.*

2) The magnetic fields used for spatial encoding are referred to as gradients because their magnitudes are engineered to vary in a linear form along any desired direction in space according to the direction of the gradient vector ($\vec{g}$). In practice, three separate gradient coils are built onto the surface of a cylinder oriented in the z-direction and these are used for generating gradient fields along each one of the three laboratory frame

cartesian directions; by energizing the gradient coils the magnitude and direction of the resulting gradient can be controlled as desired.

$$\vec{B}^{(\text{enc})}(\vec{x}, t) = \vec{x} \cdot \vec{g}(t)\hat{z} \qquad \text{(Eq. C1.4–2)}$$

Altogether the total theoretically intended magnetic field to be applied to the patient for generating images with an MRI scan is

$$\vec{B}^{(\text{Theo})}(\vec{x}, t) = B_0 \, \widehat{z} + B_1(t)(\cos(\omega_{\text{rf}} t)\hat{x} + \sin(\omega_{\text{rf}} t)\hat{y}) + \vec{x} \cdot \vec{g}(t)\hat{z}$$

$$\text{(Eq. C1.4–3)}$$

The waveform graphs of $B_1(t), g_x(t), g_y(t), g_z(t)$ as functions of time constitute the timing diagram of an MRI pulse sequence.

In practice, the actual magnetic field realized at any given $^1$H-proton position inside the patient during MRI experimentation --hereafter the experimental MRI field — will deviate from the intended theoretical field. Such deviations can stem from instrumental imperfections and most importantly, result from magnetic interactions between patient and MRI scanner.

$$\vec{B}^{(\text{Exp})}(\vec{x}, t) = \vec{B}^{(\text{Theo})}(\vec{x}, t) + \delta\vec{B}_0(\vec{x}) + \delta\vec{B}_1^{(\text{rf})}(\vec{x}, t) + \delta\vec{B}^{(\text{enc})}(\vec{x}, t)$$

$$\text{(Eq. C1.4–4)}$$

The deviations of the polarization field stem from three different sources: 1) the magnetic field inhomogeneities arising from coil and/or shimming imperfections. 2) The magnetic field inhomogeneities that arise from magnetic susceptibilities differences between tissues and structures inside the patient, and 3) the local microscopic field experienced by each $^1$H-proton is affected by the shielding effect of the electronic cloud, a phenomenon referred to as chemical shift. Accordingly, we write:

$$\delta\vec{B}_0(\vec{x}) = \delta\vec{B}_0^{(\text{coil})}(\vec{x}) + \delta\vec{B}_0^{(\chi)}(\vec{x}) + \delta\vec{B}_0^{(\text{CS})}(\vec{x}) \qquad \text{(Eq. C1.4–5)}$$

The magnetic susceptibility term can be decomposed into terms representing local --high spatial frequencies-- and global --low spatial frequencies-- variations of the magnetic field and this has implications in the field of susceptibility weighted MRI (Haacke, Mittal, Wu, Neelavalli, & Cheng, 2009).

$$\delta\vec{B}_0^{(\chi)}(\vec{x}) = \delta\vec{B}_0^{(\text{local geom})}(\vec{x}) + \delta\vec{B}_0^{(\text{global geom})}(\vec{x}) \qquad \text{(Eq. C1.4–6)}$$

In turn, the chemical shift term represents the magnetic shielding effects of the electron cloud as it reacts to the main magnetic field. These effects are of microscopic origin and exhibit very weak spatial dependences because these depend primarily on chemical structure --*i.e.* water *vs.* fat-- and not on geometry. The chemical shift term is proportional to the polarization field *via* the chemical shift constant:

$$\delta\vec{B}_0^{(CS)}(\vec{x}) = -\sigma B_0 \qquad \text{(Eq. C1.4–7)}$$

The deviations of the active $B_1$ field arise from eddy currents, inevitable imperfections in rf coil designs, and also from electrical currents induced in the patient. The latter leads to the phenomenon of dielectric resonances (Ibrahim, Lee, Abduljalil, Baertlein, & Robitaille, 2001), an adverse effect the severity of which increases at higher field strengths (3T and above) particularly for whole body applications. In MRI, the presence of dielectric resonances implies a field distribution that is a function of the shape and dielectric properties of the body only.

Finally, the deviations in the encoding gradient field can arise from nonlinearities secondary to coil design limitations and from susceptibility differences at some tissue interfaces.

Another magnetic field, which is not externally applied but is intrinsic to the patient, plays a very important role in MRI specifically as the main agent of magnetization relaxation, as will be studied in detail later (see Chapter D). In tissue, the [1]H-proton-bearing molecules are in permanent state of thermal agitation. In particular, water molecules rotate and perform random walks. As a direct consequence of these random motions, the magnetic field experienced by [1]H-protons also includes local internal magnetic fields, which change randomly in direction and amplitude as a function of time. To gain insight into the magnitude of such internal fields, we consider a magnetic moment $\vec{\mu}$ that is located at the origin of the frame of reference; according to the theory of electromagnetism, the dipolar field is:

$$\vec{B}^{(Int)}(\vec{x}) = -\frac{\mu_0}{4\pi}\left(\frac{\vec{\mu}}{|\vec{x}^3|} - \frac{3\vec{x}(\vec{x}\cdot\vec{\mu})}{|\vec{x}|^5}\right) \qquad \text{(Eq. C1.4–8)}$$

Using this formula, we calculate the magnitude of the field generated by [1]H-proton at one water inter-proton distance (*i.e.* d~1.5Å) to be approximately 0.84 mT.

## C2. The Quantum Physics Scale: Magnetic Moments

### C2.1. *Elements of Quantum Mechanics*

C2.1.i. *The Language of Quantum Theory*

Quantum mechanics is the theoretical framework within which it has been found possible to describe with accuracy the physical systems of dimensions about that of molecules and atoms, as well as subatomic particles including electrons, protons, neutrons, and many others. The mathematical formalism of quantum theory is markedly different from that of Newtonian classical mechanics. In classical mechanics, the description of a system is exhausted when its position in space and all its dynamic variables −*e.g.* linear momentum, angular momentum, and energy-- have been determined as functions of time, either analytically or by numerical computation. In quantum mechanics on the other hand, a physical system is represented by separate and different mathematical entities, specifically by state vectors –denoted as is customary by the symbol $|\psi(t)\rangle$ and $\langle\psi(t)|$ for the complex conjugate transposed vector — and operators that operate on state vectors. Such operators are customarily denoted by bold symbols, for example $\vec{\mathbf{p}}$ for the linear momentum operator. Most importantly, neither the state vectors nor the observable operators can be measured; instead, the connection with experimental information is *via* the eigenvalues of observable operators and by the probability of their occurrence in a given experiment.

C2.1.ii. *Postulates*

*Postulate 1: States of a Physical System*
The state of a physical system is completely specified by a vector of length one in a Hilbert space. Such space, which contains all possible quantum mechanical states of a system, may have infinite dimensionality depending on the system studied and is a generalization of the familiar three-dimensional Euclidean or geometric space. Specifically, Hilbert spaces are mathematical constructs consisting of a vector space that is equipped with an inner product operation that is the generalization of the more familiar dot product in geometric space. The importance of spaces equipped with an inner product is that these allow for calculating a vector's norm (length) as well as for projecting state vectors into other state vectors such as the elements of a base. Such projections will acquire a more precise meaning in the next postulate. To finalize, this postulate's discussion, state

vectors have norm one, *i.e.*

$$\langle \psi(t)|\psi(t)\rangle = 1 \qquad \text{(Eq. C2.1–1)}$$

Furthermore, any state can be expressed as a linear combination of base vectors of the Hilbert space:

$$|\psi(t)\rangle = \sum_{\text{all states}} a_n(t)|\psi_n\rangle \qquad \text{(Eq. C2.1–2)}$$

The coefficients, can be calculated by projecting the state vector onto the base elements, i.e.

$$a_n(t) = \langle \psi(t)|\psi_n\rangle \qquad \text{(Eq. C2.1–3)}$$

These coefficients, known as probability amplitudes, are in general complex numbers. Because the state has a length of one, the following equation, known as a closure equation, holds:

$$\sum_{\text{all states}} |a_n(t)|^2 = 1 \qquad \text{(Eq. C2.1–4)}$$

*Postulate 2: Observables*

To each observable in classical mechanics, there corresponds a Hermitian operator in quantum mechanics. A Hermitian operator is equal to its adjoint, *i.e.*

$$\widehat{A} = \widehat{A}^{T*} \qquad \text{(Eq. C2.1–5)}$$

The Hermitian requirement leads to real-valued eigenvalues as well as real expectation values, thus providing an intuitive physical connection with experimental results. In practice, physical observables are constructed by inference to their counterparts in classical mechanics. There are however observables in quantum mechanics that do not have a classical analog; such is the case of the intrinsic angular momentum, or spin.

The fundamental variables of position and of linear momentum of classical mechanics are represented in quantum mechanics by Hermitian operators $\hat{x}$ and $\hat{p}$ with the following matrix elements in the eigenbasis of $\hat{x}$:

$$\langle x|\hat{x}|x'\rangle = x\delta(x - x') \qquad \text{(Eq. C2.1–6)}$$

and

$$\langle x|\hat{\mathbf{p}}|x'\rangle = -i\hbar \frac{\mathrm{d}}{\mathrm{d}x}\delta(x - x') \qquad \text{(Eq. C2.1--7)}$$

where $\delta$ is the impulse function, also known as Dirac's delta function, and is heuristically defined by:

$$\delta(x) = \begin{cases} 0 & \text{if} \quad x \neq 0 \\ \infty & \text{if} \quad x = 0 \end{cases} \qquad \text{(Eq. C2.1--8)}$$

with the additional requirement that:

$$\int_{\infty}^{+\infty} \delta(x) = 1 \qquad \text{(Eq. C2.1--9)}$$

It can be shown (Shankar, 2007) that assuming the matrix elements above leads to the commutation relation:

$$[\hat{\mathbf{x}}, \hat{\mathbf{p}}] \equiv (\hat{\mathbf{x}}\hat{\mathbf{p}} - \hat{\mathbf{p}}\hat{\mathbf{x}}) = i\hbar \qquad \text{(Eq. C2.1--10)}$$

between the operators of position and linear momentum.

In 1926, Werner Heisenberg, who developed the matrix formulation of quantum mechanics, realized that the non-commuting properties between the position and linear momentum operators imply the following inequalities between the variances (see Eq. 2.1-17 below) of the operators:

$$\left. \begin{aligned} \Delta x \Delta p_x &\geq \frac{\hbar}{2} \\ \Delta y \Delta p_y &\geq \frac{\hbar}{2} \\ \Delta z \Delta p_z &\geq \frac{\hbar}{2} \end{aligned} \right\} \qquad \text{(Eq. C2.1--11)}$$

these inequalities constitute the Heisenberg uncertainty principle, which states that certain pairs of physical properties, such as position and linear momentum, cannot be simultaneously known to arbitrarily high precision.

Moreover, the position-linear momentum commutation relations above lead to the cyclic commutation relations for the components of the angular momentum operator ($\vec{\mathbf{L}} = \vec{\mathbf{r}} \times \vec{\mathbf{p}}$), specifically:

$$\left. \begin{aligned} [\mathbf{L}_x, \mathbf{L}_y] &= i\hbar\mathbf{L}_z \\ [\mathbf{L}_y, \mathbf{L}_z] &= i\hbar\mathbf{L}_x \\ [\mathbf{L}_z, \mathbf{L}_x] &= i\hbar\mathbf{L}_y \end{aligned} \right\} \qquad \text{(Eq. C2.1--12)}$$

We will see later that these commutation relations are satisfied by the components of the spin operator.

*Postulate 3: Results of Measurements*
A measurement of an observable that is represented by a generic operator $\hat{\mathbf{A}}$ will yield one of the operator's eigenvalues ($\alpha_n$), as calculated by solving the eigenvalue problem:

$$\hat{\mathbf{A}}\,|\psi_n\rangle = \alpha_n|\psi_n\rangle \qquad \text{(Eq. C2.1–13)}$$

Furthermore, the probability of obtaining a specific experimental result ($\alpha_n$) is:

$$p(\alpha_n) = |\langle\psi|\psi_n\rangle|^2 \qquad \text{(Eq. C2.1–14)}$$

The system is left in the state $|\psi_n\rangle$ after the measurement.

The probabilistic interpretation of quantum mechanics is further reinforced by defining the average value of the observable as would be obtained by repeating an experiment a large number of times:

$$\langle\hat{\mathbf{A}}\rangle = \langle\psi(t)|\,\hat{\mathbf{A}}\,|\psi(t)\rangle \qquad \text{(Eq. C2.1–15)}$$

Using postulates 2 and 3, we find:

$$\langle\hat{\mathbf{A}}\rangle = \sum_{\text{all states}} \alpha_n|a_n(t)|^2 \qquad \text{(Eq. C2.1–16)}$$

Furthermore, the variance of $\hat{\mathbf{A}}$ in this state is:

$$\Delta A \equiv \sqrt{\langle\psi|(\hat{\mathbf{A}} - \langle\hat{\mathbf{A}}\rangle)^2|\psi\rangle} \qquad \text{(Eq. C2.1–17)}$$

*Postulate 4: Equations of Motion*
In the absence of measurements or external interactions, the temporal evolution of system states is deterministic and governed by the Schrödinger equation:

$$i\hbar\frac{\partial}{\partial t}|\psi(t)\rangle = \mathbf{H}|\psi(t)\rangle \qquad \text{(Eq. C2.1–18)}$$

The energy operator $\mathbf{H}$ is known as the Hamiltonian of the system. In this particular formulation of quantum mechanics, known as the Schrödinger picture, only the system states are functions of time while the operators are time independent. Furthermore, if the Hamiltonian is constant with time, the time evolution of states is given by:

$$|\psi(t)\rangle = \exp\left(-i\frac{\mathbf{H}}{\hbar}t\right)|\psi(0)\rangle \qquad \text{(Eq. C2.1–19)}$$

Other dynamic pictures are possible; for example, it could be desirable to study a system in a picture where states are time independent and all time dependencies are in the observables. Such is the case of the Heisenberg picture, where the equation of motion of an observable is given by:

$$i\hbar\frac{d}{dt}\,\widehat{\mathbf{A}} = [\widehat{\mathbf{A}},\mathbf{H}] \qquad \text{(Eq. C2.1–20)}$$

In this equation of motion, the square bracket denotes the commutation operation $(\widehat{\mathbf{A}}\,\mathbf{H}-\mathbf{H}\,\widehat{\mathbf{A}})$ between the observable $(\widehat{\mathbf{A}})$ and the energy operator.

*Postulate 5: Intrinsic Angular Momentum (Spin) and Many Particle States*
Evidence from atomic beam experiments pointed to the fact that some experimental findings could not be explained under the assumption that the observables with classical equivalent -- specifically position $(\vec{\mathbf{x}})$, linear momentum $(\vec{\mathbf{p}})$, and angular momentum $(\vec{\mathbf{L}})$ -- did exhaust all the degrees of freedom of subatomic particles. Such experimental evidence lead to the postulation of a new observable, the intrinsic angular momentum or spin, which does not have a classical equivalent. In 1924, Wolfgang Pauli introduced the concept of a "two-valued quantum degree of freedom" associated to electrons. He later formalized in 1927 the theory of spin using quantum mechanics, as detailed in the next sections. The electron spin is an essential part of relativistic quantum mechanics.

## C2.2. *Solitary* [1]*H-Proton in a Static Magnetic Field*

We consider a single spin-1/2 particle that is immersed in a constant magnetic field in the z-direction

$$\vec{B}_0 = B_0\,\widehat{z} \qquad \text{(Eq. C2.2–1)}$$

The particle is assumed to be at rest relative to the laboratory frame of reference. The energy operator of this system is inferred from the classical energy of a classical magnetic dipole moment in an external magnetic field; hereafter the single particle Zeeman Hamiltonian is:

$$\mathbf{H}^{\text{Zeeman}} = -\vec{\mu}\cdot\vec{B}_0 \qquad \text{(Eq. C2.2–2)}$$

In turn, the magnetic moment operator is proportional to the intrinsic angular momentum or spin operator,

$$\vec{\mu} = 2\pi\gamma\vec{S} \qquad \text{(Eq. C2.2–3)}$$

Where the proportionality constant $(\gamma)$ is the gyromagnetic ratio of the particle and is a measure of how strongly the spin of a particle couples to a magnetic field. According to relativistic quantum mechanics --which is beyond the scope of this book--, the gyromagnetic ratio can be expressed in terms of the charge (q), the mass (m), and a dimensionless form factor known as the spin g-factor (g) of the particle, specifically:

$$\gamma = \frac{1}{2\pi}g\left(\frac{q}{2m}\right) \qquad \text{(Eq. C2.2–4)}$$

We are interested here in the $^{1}$H-proton ($g=5.586$, $q=1.602\ 10^{-19}$ C, and $m=1.672\ 10^{-27}$ kg), therefore with a gyromagnetic ratio of $\gamma = 42.58$ MHz/Tesla.

The proton spin g-factor is much larger than that of the electron ($g=2$), which is believed to be a point particle without internal structure. The larger spin g-factor of the proton is taken as an indication of the proton's internal structure consisting of three fundamental particles known as quarks.

According to Pauli's spin theory, for all spin-1/2 particles, including the proton, the spin operator can be represented as:

$$\vec{S} = \frac{\hbar}{2}\vec{\sigma} \qquad \text{(Eq. C2.2–5)}$$

where the three components of the vector $(\vec{\sigma})$ are the two-by-two Pauli matrices, specifically,

$$\vec{\sigma} = \sigma_x\hat{x} + \sigma_y\hat{y} + \sigma_z\hat{z} = \begin{bmatrix} 0 & 1 \\ 1 & 0 \end{bmatrix}\hat{x} + \begin{bmatrix} 0 & -i \\ i & 0 \end{bmatrix}\hat{y} + \begin{bmatrix} 1 & 0 \\ 0 & -1 \end{bmatrix}\hat{z}$$

$$\text{(Eq. C2.2–6)}$$

The reader can easily check that this spin operator satisfies the cyclic angular momentum commutation relations *i.e.*

$$\left.\begin{aligned}
[\mathbf{S_x}, \mathbf{S_y}] &= i\hbar \mathbf{S_z} \\
[\mathbf{S_y}, \mathbf{S_z}] &= i\hbar \mathbf{S_x} \\
[\mathbf{S_z}, \mathbf{S_x}] &= i\hbar \mathbf{S_y}
\end{aligned}\right\}$$
(Eq. C2.2–7)

Hence the Zeeman Hamiltonian for a spin in a constant magnetic field in the z-direction is:

$$\mathbf{H}^{\text{Zeeman}} = -\pi\hbar\gamma B_0 \sigma_z$$
(Eq. C2.2–8)

This Hamiltonian can be easily diagonalized, as explained in the following section.

### C2.2.i. *Stationary States: Longitudinal Zeeman States*

Using elemental linear algebra, the eigenvalues and eigenvectors of the Zeeman Hamiltonian can be calculated, leading to:

$$\mathbf{H}^{\text{Zeeman}} \xrightarrow{\;eigenvalues\;} \{E_{\text{up}} = -\pi\hbar\gamma B_0, E_{\text{down}} = \pi\hbar\gamma B_0\}$$
(Eq. C2.2–9)

and,

$$\mathbf{H}^{\text{Zeeman}} \xrightarrow{\;eigenvalues\;} \left\{|\text{up}\rangle = \begin{bmatrix} 1 \\ 0 \end{bmatrix} \text{and} |\text{down}\rangle = \begin{bmatrix} 0 \\ 1 \end{bmatrix}\right\}$$
(Eq. C2.2–10)

These spin-up and spin-down unitary vectors form the Zeeman base of eigenstates. Furthermore, as a consistency check, we note that the energy operator applied to these spin-up and spin-down states give the expected eigenvalue results:

$$\mathbf{H}^{\text{Zeeman}}|\text{up}\rangle = -\pi\hbar\gamma B_0 \begin{bmatrix} 1 & 0 \\ 0 & -1 \end{bmatrix} \begin{pmatrix} 1 \\ 0 \end{pmatrix} = -\pi\hbar\gamma B_0 \begin{pmatrix} 1 \\ 0 \end{pmatrix} = -\pi\hbar\gamma B_0|\text{up}\rangle$$
(Eq. C2.2–11)

and

$$\mathbf{H}^{\text{Zeeman}}|\text{down}\rangle = -\pi\hbar\gamma B_0 \begin{bmatrix} 1 & 0 \\ 0 & -1 \end{bmatrix} \begin{pmatrix} 0 \\ 1 \end{pmatrix} = -\pi\hbar\gamma B_0 \begin{pmatrix} 0 \\ -1 \end{pmatrix}$$
$$= +\pi\hbar\gamma B_0|\text{down}\rangle$$
(Eq. C2.2–12)

From the equations above, the resonant frequency of excitation is proportional to energy difference between eigenstates:

$$\omega_0 \equiv \frac{(E_{down} - E_{up})}{\hbar} = 2\pi\gamma B_0 \qquad \text{(Eq. C2.2–13)}$$

Furthermore, for these stationary states, which have well-defined energy values, the expectation values of the magnetic moment vector are:

$$\langle up|\vec{\mu}|up\rangle = \begin{bmatrix} \langle \mu_x \rangle = 0 \\ \langle \mu_y \rangle = 0 \\ \langle \mu_z \rangle = \pi\hbar\gamma \end{bmatrix} = +\mu_{Proton}\,\hat{z} \qquad \text{(Eq. C2.2–14)}$$

and,

$$\langle down|\vec{\mu}|down\rangle = \begin{bmatrix} \langle \mu_x \rangle = 0 \\ \langle \mu_y \rangle = 0 \\ \langle \mu_z \rangle = -\pi\hbar\gamma \end{bmatrix} = -\mu_{Proton}\,\hat{z}$$

$$\text{(Eq. C2.2–15)}$$

Hence, the natural definition for the magnitude of the proton's magnetic moment is:

$$\mu_{Proton} \equiv \pi\hbar\gamma \qquad \text{(Eq. C2.2–16)}$$

Accordingly, the Zeeman eigenstates represent a magnetic moment that on average is oriented parallel (spin up) or anti-parallel (spin down) to the applied magnetic field and have null average components in the x-y transverse plane. Therefore, these stationary states are referred to as longitudinal magnetic moment states.

C2.2.ii. *Mixture States: Transverse Magnetic Dipole Moments*

We study the temporal evolution of a mixture spin state, which in its most general form is a linear combination of the spin-up and spin-down states, specifically:

$$|\psi(t)\rangle = a_{up}(t)|up\rangle + a_{down}(t)|down\rangle \qquad \text{(Eq. C2.2–17)}$$

Alternatively, such mixture spin state can be expressed as a vector in a two-dimensional space with the spin-up and spin-down probability amplitudes as components, *i.e.*

$$|\Psi(t)\rangle = \begin{bmatrix} a_{up}(t) \\ a_{down}(t) \end{bmatrix} \quad \text{and} \quad \langle\Psi(t)| = [a_{up}^*(t), a_{down}^*(t)]$$

(Eq. C2.2–18)

Remembering that the spin operator in this space assumes the matrix form,

$$\vec{S} = S_x\hat{x} + S_y\hat{y} + S_z\hat{z} = \frac{\hbar}{2}\left\{ \begin{bmatrix} 0 & 1 \\ 1 & 0 \end{bmatrix}\hat{x} + \begin{bmatrix} 0 & -i \\ i & 0 \end{bmatrix}\hat{y} + \begin{bmatrix} 1 & 0 \\ 0 & -1 \end{bmatrix}\hat{z} \right\}$$

(Eq. C2.2–19)

We proceed to calculate the time dependent expectation value of the magnetic dipole moment operator, *i.e.*

$$\vec{\mu}(t) = \langle\psi(t)|\vec{\mu}|\psi(t)\rangle$$

(Eq. C2.2–20)

Using the equations above, the magnetic moment's expectation value is given by:

$$\vec{\mu}(t) = \mu_{Proton}\begin{pmatrix} 2\,\text{Re}\{a_{up}^*(t)a_{down}(t)\} \\ 2\,\text{Im}\{a_{up}^*(t)a_{down}(t)\} \\ |a_{up}(t)|^2 - |a_{down}(t)|^2 \end{pmatrix}$$

(Eq. C2.2–21)

Hence, the x- and the y-components of the magnetic moment depend on the mixed (product) spin-up and spin-down probabilities. Furthermore, these components can be naturally combined to form a transverse magnetic moment phasor ($\mu_T \equiv \mu_x + i\mu_y$), hence:

$$\mu_T(t) = 2\mu_{Proton}\,a_{up}^*(t)a_{down}(t)$$

(Eq. C2.2–22)

Remembering the normalization condition ($|a_{up}(t)|^2 + |a_{down}(t)|^2 = 1$), the z-component --hereafter the longitudinal component-- of the magnetic moment expectation value reads,

$$\mu_z(t) = \mu_{Proton}(1 - 2|a_{down}(t)|^2)$$

(Eq. C2.2–23)

Note that the last two equations above are very general and also apply to a spin in a time dependent magnetic field, provided that the dynamic problem is formulated using the Zeeman eigenstates.

The energy of a spin in a mixture state is:

$$E(t) = \langle \psi(t) | \mathbf{H}^{\text{Zeeman}} | \psi(t) \rangle = E_{\text{up}}(1 - 2|a_{\text{down}}(t)|^2)$$

$$\text{(Eq. C2.2-24)}$$

We note that the spin-up and spin-down probability amplitudes as functions of time are given by:

$$a_{\text{up,down}}(t) = a_{\text{up,down}}(0) \exp \left[ -i \frac{E_{\text{up,down}}}{\hbar} t \right] \quad \text{(Eq. C2.2-25)}$$

Remembering that the definition of the Larmor frequency of excitation ($\omega_0 \equiv (E_{\text{down}} - E_{\text{up}})/\hbar = 2\pi\gamma B_0$), we therefore find that for a spin-1/2 particle in a mixture state, the expectation values of the transverse and longitudinal magnetic moment components are:

$$\mu_T(t) = 2\mu_{\text{Proton}} a_{\text{up}}^*(0) a_{\text{down}}(0) \exp[-i\omega_0 t] \quad \text{(Eq. C2.2-26)}$$

and

$$\mu_z(t) = \mu_{\text{Proton}}(1 - 2|a_{\text{down}}(0)|^2) = \mu_z(0) \quad \text{(Eq. C2.2-27)}$$

Furthermore, the energy is given by

$$E(t) = E_{\text{up}}(1 - 2|a_{\text{down}}(t)|^2) = E_{\text{up}}(1 - 2|a_{\text{down}}(0)|^2) = E(0)$$

$$\text{(Eq. C2.2-28)}$$

Hence, the z-component of the expectation value as well as the energy are conserved (time independent) quantities.

In summary, the expectation value magnetic moment is a vector that moves on the surface of a cone centered about the z-axis such that the transverse component rotates clockwise with a frequency equal to Larmor frequency of excitation; this phenomenon is therefore known as Larmor Precession.

## C2.2.iii. *The Transverse Spin State*

As a very important example, let us consider a mixture state that was prepared at time zero such that the spin-up and spin-down amplitudes were equal:

$$|a_{up}(0)|^2 = |a_{down}(0)|^2 = \frac{1}{2} \qquad \text{(Eq. C2.2–29)}$$

We will see later on that such states can be generated by applying a time varying radiofrequency magnetic field of the appropriate amplitude and duration. For the time being, we note that when inserting these initial amplitudes in Eq. C2.2-26–28, we find:

$$\mu_T(t) = \mu_{Proton} \exp[-i\omega_0 t] \qquad \text{(Eq. C2.2–30)}$$

and

$$\mu_z(t) = 0 \qquad \text{(Eq. C2.2–31)}$$

In addition, the energy is constant, specifically:

$$E(t) = 0 \qquad \text{(Eq. C2.2–32)}$$

These results describe a magnetic moment that on average precesses in the transverse plane at the Larmor frequency. Furthermore, because the energy is constant, the system will remain in this state permanently unless disturbed.

### C2.3. *Solitary $^1$H-Proton in a Time Dependent Magnetic Field*

C2.3.i. *The Interaction Schrödinger Equation*

We study the spin dynamics of a solitary $^1$H-proton in a magnetic field of the form:

$$\vec{B}(t) = B_0 \, \widehat{z} + \vec{B}^{Int}(t) \qquad \text{(Eq. C2.3–1)}$$

Where the second term denotes an arbitrary time-dependent magnetic field intended to model either the applied magnetic field pulses used in MRI experimentation or the randomly fluctuating interaction magnetic fields that $^1$H-protons experience in tissue.

We begin by solving formally the corresponding interaction Schrödinger equation, specifically:

$$i\hbar \frac{\partial}{\partial t}|\psi(t)\rangle = -2\pi\gamma\vec{S}\cdot[(B_0 + B_z^{(Int)}(t))\hat{z} + (B_x^{(Int)}(t)\hat{z} + B_y^{(Int)}(t)\hat{y})]|\psi(t)\rangle$$

$$\text{(Eq. C2.3–2)}$$

We formulate the problem in the Zeeman basis of eigenvectors and use the symbols $(b_{up}(t), b_{down}(t))$ for designating the spin-up and spin-down probability amplitudes; accordingly:

$$|\psi(t)\rangle = b_{up}(t)|up\rangle + b_{down}(t)|down\rangle \qquad \text{(Eq. C2.3–3)}$$

As before, the state normalization condition reads:

$$|b_{up}(t)|^2 + |b_{down}(t)|^2 = 1 \qquad \text{(Eq. C2.3–4)}$$

The new spin-up and spin-down probability amplitudes must be such that the spin state $(|\psi(t)\rangle)$ satisfies the Schrödinger equation with time dependent magnetic interaction terms above. We can easily prove that this is equivalent to solving the following system of coupled differential equations for the spin-up and spin-down probability amplitudes:

$$\frac{db_{up}}{dt} = +i\pi\gamma(B_0 + B_z^{(Int)}(t))b_{up} + i\pi\gamma B_T^{(Int)*}(t)b_{down}$$

$$\text{(Eq. C2.3–5)}$$

and

$$\frac{db_{down}}{dt} = -i\pi\gamma(B_0 + B_z^{(Int)}(t))b_{down} + i\pi\gamma B_T^{(Int)}(t)b_{up}$$

$$\text{(Eq. C2.3–6)}$$

In the equations above, we have defined the interaction transverse magnetic field phasor:

$$B_T^{(Int)}(t) \equiv B_x^{(Int)}(t) + iB_y^{(Int)}(t) \qquad \text{(Eq. C2.3–7)}$$

which will allows writing the equations of motion in a form particularly suitable for transitioning to the spatial scale of spin packets.

### C2.3.ii. *Magnetic Moment Dynamics: Differential and Integral Equations of Motion*

As pointed out in the previous section, the formulas for the transverse and longitudinal components of magnetic moment expectation value $(\vec{\mu}(t) = \langle\psi(t)|\vec{\mu}|\psi(t)\rangle)$ as functions of the Zeeman spin-up and spin-down probability amplitudes do not depend on the specific Hamiltonian and therefore these are unchanged (see Eq. C2.2–19, 20). Hence, we have:

$$\mu_T(t) = 2\mu_{Proton}b^*_{up}(t)b_{down}(t) \qquad (Eq.\ C2.3\text{--}8)$$

and

$$\mu_z(t) = \mu_{Proton}(1 - 2|b_{down}(t)|^2) \qquad (Eq.\ C2.3\text{--}9)$$

By taking the time derivatives of these equations and using the intermediate results of the previous section (Eq. C2.3-5, 6), we find the following set of coupled differential equations that govern the temporal evolution of the magnetic moment's expectation value:

$$\frac{d\mu_T}{dt} = -i2\pi\gamma(B_0 + B_z^{(Int)}(t))\mu_T(t) + i2\pi\gamma B_T^{(Int)}(t)\mu_z(t)$$

$$(Eq.\ C2.3\text{--}10)$$

and

$$\frac{d\mu_z}{dt} = -2\pi\gamma\,\mathrm{Im}\{B_T^{(Int)}(t)^*\mu_T(t)\} \qquad (Eq.\ C2.3\text{--}11)$$

These two differential equations are the quantum mechanical precursors of the Bloch equation(s). In particular, we note that the last equation can be written in several equivalent forms:

$$\frac{d\mu_z}{dt} = -2\pi\gamma\,\mathrm{Im}\{B_T^{(Int)}(t)^*\mu_T(t)\} = 2\pi\gamma(\vec{\mu} \times \vec{B}^{(Int)})_z$$

$$(Eq.\ C2.3\text{--}12)$$

which as we will see later, appear normally in the context of the semi-classical Bloch equation.

C2.3.iii. *Integral Equations*

The differential equations above can be integrated formally resulting in the following set of coupled integral equations for the transverse and longitudinal components of the magnetic moment expectation values, these are:

$$\mu_T(t) = \left\{\mu_T(0) + i2\pi\gamma \int_0^t B_T^{(Int)}(t')\mu_z(t')\exp[+i(\omega_0 t' + \varphi_F(t'))]dt'\right\}$$

$$\times \exp[-i(\omega_0 t + \varphi_F(t))] \qquad (Eq.\ C2.3\text{--}13)$$

and,

$$\mu_z(t) = \mu_z(0) + i\pi\gamma \int_0^t (B_T^{(Int)}(t')^*\mu_T(t') - B_T^{(Int)}(t')\mu_T^*(t'))dt'$$

$$\text{(Eq. C2.3–14)}$$

Where the accrued phase, specifically:

$$\varphi_F(t) = 2\pi\gamma \int_o^t B_z^{(Int)}(t')dt' \qquad \text{(Eq. C2.3–15)}$$

is caused by fluctuations of the precession frequency. In turn, these precession frequency fluctuations result from the spin-down eigenstate energy fluctuating in time thus causing a flutter effect of the Larmor frequency. In tissue, each separate spin experiences different and uncorrelated perturbation (dephasing) fields and this mechanism is main cause of the macroscopic phenomenon of transverse magnetization decay *via* dephasing, as will be studied later in the book. The correlation time is defined as the time interval during which the local magnetic field does not change abruptly for example, because of molecular tumbling and collisions.

In summary, the following interpretation for each the terms of the transverse equation of motion are:

$$\underbrace{\frac{d\mu_T}{dt}} = \underbrace{-i\omega_0\mu_T(t)}_{\substack{\text{On-resonance precession} \\ \downarrow \\ \text{Free precession}}} \underbrace{-i2\pi\gamma B_z^{(Int)}(t)\mu_T(t)}_{\substack{B_z^{(Int)}\text{-Flutter} \\ \downarrow \\ \text{Secular-}T_2}} \underbrace{+i2\pi\gamma B_T^{(Int)}(t)\mu_z(t)}_{\substack{B_T^{(Int)}\text{-Flip-flopping} \\ \downarrow \\ T_1\text{-contribution to }T_2}}$$

$$\text{(Eq. C2.3–16)}$$

The first term on the right hand side of this equation describes precession with constant angular frequency, or Larmor precession. The second term describes precession with a time-varying angular frequency thus altering the phase of the transverse magnetic moment. This will later lead on to the description of spin packet dephasing, which in turn leads to the definition of the secular component of the transverse relaxation time $T_2$. Finally, the third term, describes transitions between the spin-up and spin-down states thus representing processes by which the transverse state is destroyed. When describing spin packets in the following sections, this term will be instrumental for defining the non-secular- or $T_1$-contibution to $T_2$.

## C3. The Semi-Classical NMR Physics Scale: Magnetized Spin Packets

### C3.1. *Nuclear Magnetization of Spin Packets*

The protonic magnetization --simply referred to as magnetization for brevity-- of a spin packet is a semi-classical physical quantity that represents the collective magnetic response of a numerous group of $^1$H-proton spins. The number of spins of a spin packet is assumed sufficiently large so the laws of statistic apply. Mathematically, the magnetization is defined as the vector sum of the magnetic moment expectation values for all $^1$H-protons contained inside an elemental packet volume ($\delta V_p$), specifically:

$$\vec{M}(\vec{x}, t) \equiv \sum_{\vec{\mu}^{(m)} \subseteq \delta V_p(\vec{x})} \vec{\mu}^{(m)}(t) \qquad \text{(Eq. C3.1–1)}$$

The dependence of the magnetization on the continuous position vector is shown explicitly in the range of the packet summation symbol. As such, the magnetization is a continuous function of space and time.

The complex transverse magnetization is given by:

$$m(\vec{x}, t) = \sum_{\vec{\mu}^{(m)} \subseteq \delta V_p(\vec{x})} \mu_T^{(m)}(t) \qquad \text{(Eq. C3.1–2)}$$

and for the longitudinal magnetization:

$$M_z(\vec{x}, t) = \sum_{\vec{\mu}^{(m)} \subseteq \delta V_p(\vec{x})} \mu_z^{(m)}(t) \qquad \text{(Eq. C3.1–3)}$$

Analogously, for the two components in the transverse plane:

$$M_{x,y}(\vec{x}, t) = \sum_{\vec{\mu}^{(m)} \subseteq \delta V_p(\vec{x})} \mu_{x,y}^{(m)}(t) \qquad \text{(Eq. C3.1–4)}$$

Hence, the complex transverse magnetization also reads:

$$m(\vec{x}, t) = M_x(\vec{x}, t) + iM_y(\vec{x}, t) \qquad \text{(Eq. C3.1–5)}$$

NMR signals generated with the nuclear induction method are modeled as volume integrals of the transverse magnetization:

$$S(t) = \iiint_{\left\{ \begin{smallmatrix} \text{Infinite} \\ \text{space} \end{smallmatrix} \right\}} \Omega(\vec{x}) m(\vec{x}, t) d^3x \qquad \text{(Eq. C3.1–6)}$$

where the function $\Omega(\vec{x})$ represents the volumetric sensitivity profile of the receiver coil used for signal reception. As such, NMR signals are complex valued functions of time; experimentally, the real and imaginary parts are separated by means of phase sensitive detectors.

### C3.2. *State of Thermal Equilibrium: Statistical Mechanics*

Even though a protonic spin packet is an ensemble of identical fermions, specifically $^1$H-protons, because the inter-proton separations in all solid state matter including biological tissue are restricted by the electron clouds to be about five orders of magnitude larger ($\sim 10^{-10}$m) than that of the proton's wavelength ($\sim 10^{-15}$m), the statistical quantum mechanical effects of the Fermi distribution can be neglected. Therefore, the numbers of spins per energy state obey a classical Maxwell-Boltzmann distribution.

Accordingly, for a spin packet consisting of PD protons per unit volume, the fractional densities of particles occupying the spin-up and the spin-down eigenstates are given respectively by

$$n_{up,down}^{(eq)}(\vec{x}, T) = \left( \frac{\exp(-E_{up,down}/2k_BT)}{\exp(-E_{up}/2k_BT) + \exp(-E_{down}/2k_BT)} \right) PD(\vec{x}, T)$$

(Eq. C3.2–1)

where T is the temperature, $k_B$ is Boltzmann's constant, and the denominator is known as the partition function of the spin ensemble.

Furthermore, the longitudinal magnetization is:

$$M_z^{(eq)}(T) = \mu_{Proton}(n_{up}^{(eq)} - n_{down}^{(eq)})$$

(Eq. C3.2–2)

and remembering that $E_{up} = -E_{down} = -\pi\hbar\gamma B_0$, then the thermal equilibrium z-magnetization, or polarization of a spin packet as a function of temperature is given by:

$$M_z^{(eq)}(\vec{x}, T) = \mu_{Proton}PD(\vec{x}, T)\tanh\left( \frac{\pi\hbar\gamma B_0}{2k_BT} \right)$$

(Eq. C3.2–3)

At room temperature (T$=300°$K) and for polarizing field strengths in the range of interest here (*i.e.* 0.5–12T), the Zeeman energies ($10^{-4}$eV to $10^{-5}$eV) are several orders of magnitude smaller than the thermal energy (0.05 eV). Therefore ($E_{up,down}/k_BT \ll 1$) and the hyperbolic tangent function can be approximated by Taylor's series first term (*i.e.* $\tanh(x) \approx x$) and the following equation, which holds to a very high degree of accuracy, is commonly used:

$$M_z^{(eq)}(\vec{x}, T) = \left(\frac{\pi^2 \gamma^2 \hbar^2 B_0}{k_B T}\right) PD(\vec{x}, T) \qquad \text{(Eq. C3.2–4)}$$

In order to gain insight about the state of thermal equilibrium of the transverse magnetization, we rewrite the single particle result of the previous chapter by labeling it with the particle index "m", specifically:

$$\mu_T^{(m)}(t) \cong \mu_T^{(m)}(0) \exp[-i(\omega_0 t + \varphi_F^{(m)}(t))] \qquad \text{(Eq. C3.2–5)}$$

Then, the spin packet sum reads:

$$m(\vec{x}, t) \cong \exp[-i\omega_0 t] \sum_{\vec{\mu}^{(m)} \subseteq \delta V_p(\vec{x})} \mu_T^{(m)}(0) \exp[-i\varphi_F^{(m)}(t)]$$

$$\text{(Eq. C3.2–6)}$$

clearly showing that in thermal equilibrium and for times long compared to the correlation time, the spin packet sum of random phase terms leads to a near null complex transverse magnetization, *i.e.*

$$m^{(eq)}(\vec{x}, t) \cong 0 \qquad \text{(Eq. C3.2–7)}$$

Hence, the thermal equilibrium magnetization vector of a spin packet is parallel to the polarizing field and given by

$$\vec{M}^{(eq)}(\vec{x}, T) = \begin{pmatrix} M_x^{(eq)} \\ M_y^{(eq)} \\ M_z^{(eq)} \end{pmatrix} \cong \begin{pmatrix} 0 \\ 0 \\ \left(\dfrac{\pi^2 \gamma^2 \hbar^2 B_0}{k_B T}\right) PD(\vec{x}, T) \end{pmatrix}$$

$$\text{(Eq. C3.2–8)}$$

In this equation, we have used the symbol "$\cong$" to indicate equality in a temporo-statistical sense, and not an exact identity in the strict mathematical sense. In other words, the magnetization vector in thermal equilibrium is primarily oriented parallel to the z-axis but it is also performing a wobbling motion around it.

### C3.3. *NMR Dynamics: The Bloch Equation*

Besides the state of thermal equilibrium described above, spin packets can also be in non-equilibrium magnetization states, which are inherently unstable. Such away from equilibrium magnetization states will exist transiently as a patient is placed inside an MRI scanner and becomes longitudinally polarized, a process that takes about twenty seconds at 1.5Tesla to reach full polarization in all tissues. Moreover, once a patient has been fully polarized, the magnetization can be removed from equilibrium *via* resonant magnetization manipulation with oscillatory $B_1$ fields.

The fundamental dynamical cycle of the magnetization during MR experimentation can be represented by:

$$\vec{M}^{(eq)}(\vec{x}) \xrightarrow{\vec{B}_1^{(rf)}} \vec{M}(\vec{x}, t) = \begin{pmatrix} M_x(\vec{x}, t) \\ M_y(\vec{x}, t) \\ M_z(\vec{x}, t) \end{pmatrix} \begin{matrix} \xrightarrow{T_2} \\ \xrightarrow{T_2} \\ \xrightarrow{T_1} \end{matrix} \begin{pmatrix} 0 \\ 0 \\ M_z^{(eq)} \end{pmatrix} = \vec{M}_z^{(eq)}(\vec{x})$$

(Eq. C3.3–1)

In this particular case, the state of the thermal equilibrium is reestablished following full relaxation of all three components.

The key features of away from equilibrium magnetization states are that $\vec{M}(\vec{x}, t)$ can be oriented in any direction and that the magnitude will not exceed that of the equilibrium magnetization at any time, *i.e.*

$$|\vec{M}(\vec{x}, t)| \leq M_z^{(eq)}(\vec{x}) \qquad \text{(Eq. C3.3–2)}$$

Furthermore, in the absence of an applied rf field, $\vec{M}(\vec{x}, t)$ will always evolve in time towards reestablishing the state of thermal equilibrium.

NMR dynamics is the theoretical framework for describing the temporal evolution of the magnetization under the influence of arbitrary magnetic fields. Of practical importance is the study of magnetization dynamics for the following experimental conditions:

a) Effects of pulsed oscillatory magnetic fields of duration much shorter than the relaxation time scale commonly referred to as transient phenomena.
b) Free relaxation in the absence of externally applied oscillatory magnetic fields.
c) Relaxation in the presence of an oscillatory magnetic field pulse of long duration: such is the case of spin locking pulse sequences. This leads to different relaxation tissue properties, namely $T_{1\rho}$ and $T_{2\rho}$.

In general, dynamical theories are built upon equations of motion; in the case of NMR dynamics *Felix Bloch* proposed in 1946 the following equation motion (Bloch, 1946), which bears his name:

$$\frac{d\vec{M}}{dt} = 2\pi\gamma\vec{M} \times \vec{B}^{(\text{Exp})} + \frac{(M_z^{(\text{eq})} - M_z)}{T_1}\hat{z} - \frac{M_x\hat{x} + M_y\hat{y}}{T_2}$$

$$\text{(Eq. C3.3–3)}$$

The first term of the Bloch equation, which can be derived easily and transparently using quantum NMR theory, represents the magnetic torques exerted by the applied or experimental magnetic fields that act on the magnetization at any given time. The magnetic fields of this magnetic torque term include the intended theoretical fields for generating and manipulating the magnetization during NMR or MRI experimentation, as well as the inevitable and unwanted deviations from such intended fields; as discussed earlier (see section C1.4) these stem from hardware imperfections and patient-scanner interactions.

The magnetic torque term does not include the effects of the randomly fluctuating dipolar magnetic fields that are present at the microscopic scale and that are generated by neighboring magnetic dipoles, specifically other $^1$H-protons and possibly unpaired orbital electrons if paramagnetic solutes are present. The net macroscopic effect of these local microscopic dipolar fields is thermal magnetization equilibration, or NMR relaxation. The relaxation processes are represented in Bloch's semi-classical theory by two relaxation times; the longitudinal magnetization recovery time ($T_1$) provides a measure of stability of the away from equilibrium z-component of the magnetization and the transverse magnetization decay time ($T_2$) provides a measure of stability of the away from equilibrium x-, y-magnetization components. The exponential nature of the relaxation processes follows directly from the form of the Bloch relaxation terms.

One should bear in mind, that the Bloch equation as written above implicitly assumes a vanishingly small state of equilibrium in the transverse plane ($M_x^{(\text{eq})} = M_y^{(\text{eq})} \cong 0$). As discussed in the previous section, this is true in a temporo-statistical sense; it is not however an exact mathematical equality and the symmetry of the relaxations terms can be better appreciated when the three components of the equilibrium magnetization are made explicit, *i.e.*

$$\frac{d\vec{M}}{dt} = 2\pi\gamma\vec{M} \times \vec{B}^{(\text{Exp})} + \frac{(M_z^{(\text{eq})} - M_z)}{T_1}\hat{z}$$

$$+ \frac{(M_x^{(\text{eq})} - M_x)\hat{x} + (M_y^{(\text{eq})} - M_y)\hat{y}}{T_2} \qquad \text{(Eq. C3.3–4)}$$

This form of the Bloch equation, which is more symmetric, shows explicitly the mathematical similarities among the three relaxation terms.

The relaxation terms of the Bloch equation are phenomenological because the relaxation times as well as the values of the thermal equilibrium magnetization are external parameters to this semi-classical theory. Insight into the microscopic properties of tissues is gained by theoretical modeling using quantum theory of relaxation (see Chapter D), thus justifying the semi-classical denomination of Bloch's theory. In practice, these three primary MR parameters either are the objects of measurement at a global spatial scale with qNMR or spatially resolved at the voxel scale *via* qMRI mapping.

### C3.4. *The Transverse and Longitudinal Bloch Equations*

The Bloch equation, which was written above in vector form, can be rewritten as a system of three coupled differential equations; one each for the x-, y-, z- cartesian components of the magnetization vector, specifically:

$$\frac{dM_x}{dt} = 2\pi\gamma(\vec{M} \times \vec{B}^{(Exp)})_x - \frac{M_x}{T_2} \qquad \text{(Eq. C3.4–1)}$$

and,

$$\frac{dM_y}{dt} = 2\pi\gamma(\vec{M} \times \vec{B}^{(Exp)})_y - \frac{M_y}{T_2} \qquad \text{(Eq. C3.4–2)}$$

and,

$$\frac{dM_z}{dt} = 2\pi\gamma(\vec{M} \times \vec{B}^{(Exp)})_z + \frac{(M_z^{(eq)} - M_z)}{T_1} \qquad \text{(Eq. C3.4–3)}$$

Alternatively and equivalently, these can be written as a system of two coupled differential equations: one for the complex-valued transverse magnetization phasor and one for real-valued longitudinal component of the magnetization. We will refer to these as the transverse and longitudinal Bloch equations, which as the reader can easily verify, are:

$$\frac{dm}{dt} = -i2\pi\gamma B_z^{(Exp)}m - \frac{m}{T_2} + i2\pi\gamma B_{1T}^{(Exp)}M_z \qquad \text{(Eq. C3.4–4)}$$

and,

$$\frac{dM_z}{dt} = -2\pi\gamma Im\{B_{1T}^{(Exp)*}m\} + \frac{(M_z^{(eq)} - M_z)}{T_1} \qquad \text{(Eq. C3.4–5)}$$

For these equations, the transverse $B_1$ phasor has been defined in the usual manner, that is:

$$B_{1T}^{(Exp)} \equiv B_{1x}^{(Exp)} + iB_{1y}^{(Exp)} \qquad \text{(Eq. C3.4–6)}$$

This last formulation of Bloch's problem in terms of just two coupled differential equations, specifically in terms of the transverse and longitudinal Bloch equations, is physically intuitive because it reflects the perfect dynamical symmetry that exists in the transverse plane between the x- and the y-components, as well as the high longitudinal-transverse asymmetry of the physical problem of matter polarized by magnetic field in the z-direction. In other words, the x- and y-components of the magnetization share common dynamical properties, which in turn are fundamentally different from that of that of the z-component. For these reasons, we will adopt throughout this book this last formulation of the Bloch equation instead of the more conventional formulation in terms of cartesian components. In particular, we will use this formulation for extending the Bloch problem to include the effects of micro-kinetics as well as magnetization exchange phenomena between spin pools with different kinetic properties.

### C3.5. *Bloch Theory: Assumptions and Limitations*

Bloch's semi-classical theory is based on several assumptions that can limit its range of applicability, specifically:

*Assumption 1*: The effects of spin-lattice and spin-spin interactions can be described by simple relaxation terms that are proportional to the instantaneous magnetization and inversely proportional to the relaxation times. This assumption implies that relaxation processes of longitudinal recovery and transverse magnetization decay are intrinsically exponential processes.

*Assumption 2*: Spins are assumed motionless. Extensions of Bloch's theory to include the kinetic effects of diffusion and flow are discussed in Section C3.6.

*Assumption 3*: Spin packets contain only one type of spins, which are therefore equivalent in terms of interactions with their surroundings. In other words, magnetization exchange phenomena are not present.

Extensions of Bloch's theory to include the effects of magnetization transfer are discussed in Section C3.7.

*Assumption 4*: Interaction of the spins with the lattice and with each other can be considered independently of their interaction with the externally applied magnetic field. In other words, relaxation effects take place in a time scale much longer than the duration of rf pulses: *i.e.* $T_1$, $T_2 \gg T_{rf}$. Such is the case for most MRI pulse sequences with the exception of spin locking pulse sequences, which were reviewed in Section C3.9.

## C3.6. *The Bloch-Torrey-Stejskal Equations: Kinetic Effects*

Henry C. Torrey generalized (Torrey, 1956) the phenomenological Bloch equations to include the effect of isotropic spin diffusion. Subsequently, Edward O. Stejskal (Stejskal, 1965) extended Torrey's theory for describing the cases of anisotropic diffusion, restricted diffusion, and flow. The resulting equation of motion referred to as the Bloch-Torrey-Stejskal (BTS) equation, reads:

$$\frac{\partial \vec{M}}{\partial t} = 2\pi\gamma\vec{M} \times \vec{B}^{(Exp)} + \frac{(M_z^{(eq)} - M_z)}{T_1}\hat{z} - \frac{(M_x\hat{x} + M_y\hat{y})}{T_2}$$
$$+ \vec{\nabla}\cdot\mathbf{D}\cdot\vec{\nabla}\vec{M} - \vec{v}\cdot\vec{\nabla}\vec{M} \qquad \text{(Eq. C3.6–1)}$$

The last two terms on the right hand side represent the effects on the magnetization caused by diffusion and flow respectively and these will be therefore referred to as the kinetic terms. The diffusion tensor ($\mathbf{D}$), which at each point in space and time is a 3×3 matrix, accounts for diffusion within and relative to the medium, which moves with a velocity ($\vec{v}$).

$$\mathbf{D}(\vec{x}) = \begin{pmatrix} D_{xx}(\vec{x}) & D_{xy}(\vec{x}) & D_{xz}(\vec{x}) \\ D_{yx}(\vec{x}) & D_{yy}(\vec{x}) & D_{yz}(\vec{x}) \\ D_{zx}(\vec{x}) & D_{zy}(\vec{x}) & D_{zz}(\vec{x}) \end{pmatrix} \qquad \text{(Eq. C3.6–2)}$$

The BTS equation above differs from the Bloch equation in the addition of the diffusion and flow terms; it also differs in that use of partial spatial and temporal derivatives becomes necessary to account for the spatio-temporal dependence of the magnetization. Furthermore, the diffusion tensor and the flow velocity may be functions of position and time.

As in the previous section, the transverse and longitudinal BTS equations are easily derived resulting in:

$$\frac{\partial m}{\partial t} = -i2\pi\gamma B_z^{(Exp)} m - \frac{m}{T_2} + i2\pi\gamma B_{1T}^{(Exp)} M_z$$

$$+\vec{\nabla}\cdot\mathbf{D}\cdot\vec{\nabla} m - \vec{v}\cdot\vec{\nabla} m \qquad \text{(Eq. C3.6–3)}$$

and

$$\frac{\partial M_z}{\partial t} = -2\pi\gamma Im\{B_{1T}^{(Exp)*} m\} + \frac{(M_z^{(eq)} - M_z)}{T_1}$$

$$+\vec{\nabla}\cdot\mathbf{D}\cdot\vec{\nabla}(M_z - M_z^{(eq)}) - \vec{v}\cdot\vec{\nabla}(M_z - M_z^{(eq)}) \qquad \text{(Eq. C3.6–4)}$$

Often one deals with isotropic-diffusion systems and with negligible flow effects; such systems are well described by the much simpler equations, which are commonly referred to as the Bloch-Torrey (BT) equations:

$$\frac{\partial m}{\partial t} = -i2\pi\gamma B_z^{(Exp)} m - \frac{m}{T_2} + i2\pi\gamma B_{1T}^{(Exp)} M_z + D\nabla^2 m$$

$$\text{(Eq. C3.6–5)}$$

and,

$$\frac{\partial M_z}{\partial t} = -2\pi\gamma Im\{B_{1T}^{(Exp)*} m\} + \frac{(M_z^{(eq)} - M_z)}{T_1} + D\nabla^2(M_z - M_z^{(eq)})$$

$$\text{(Eq. C3.6–6)}$$

### C3.7. *The Two-Pool Bloch-Torrey-Stejskal (BTS) Equations: Magnetization Exchange*

The preceding excitation-relaxation Bloch equations and its generalizations to include micro-and macro-kinetic phenomena *via* the diffusion and flow velocity terms, accurately describe spin packets consisting of a single spin type. More precisely, a spin type refers to spins with not only the same gyromagnetic ratio but additionally with the same average magneto-kinetic properties. For example, all the mobile water $^1$H-protons in the cytoplasm of a cell may be thought as having the same motional degrees of freedom --later to be characterized by a correlation time-- and in addition as experiencing equivalent randomly changing dipolar fields from the neighboring spins. The concept of equivalent randomly changing dipolar fields should be interpreted in terms of its net temporal-average effect over a time much longer that the time between collisions; such inter-collision time is a few picoseconds in low viscosity fluids such as water at room temperature. Other spin types in biological tissue are the less mobile $^1$H-protons of the water molecules of the hydration layers of macromolecules, as well as the $^1$H-protons of the macromolecules themselves (*e.g.* $^1$H-protons in lipids and proteins).

Owing to the high molecular diversity of biological tissue, one can logically expect that biological spin packets could consist of several spin types, in general more than two. However, at the microscopic spatial scale of spin packets, *i.e.* 6–12 nm, one can reasonably assume that only two spin types could suffice for describing the relaxation properties. As discussed above (Section C1.3), spin packets are smaller than the size of the finest sub-cellular structures (*e.g.* thinnest axons) and we can therefore safely assume that spin packets do not contain closed tissue compartments. The presence of closed tissue compartments would lead to at least three spin types, specifically intra-compartmental, extra-compartmental, and hydration spins.

For such an idealized two-pool model of spin packets, we will index all physical quantities with the superscript "A" to refer to the main spin type, which is the signal-generating spin type because it is mobile; it is often referred to as the liquid pool. The superscript "B" will be used to refer to the less mobile spin type, which is often referred to as the semisolid pool. Accordingly, the thermal equilibrium longitudinal magnetization of such two-pool spin packet is:

$$M_z^{(eq)} = M_z^{(A\text{-}eq)} + M_z^{(B\text{-}eq)} \qquad \text{(Eq. C3.7–1)}$$

and each spin pool is described by single-pool Bloch-Torrey-Stejskal differential equations modified by inter-pool magnetization exchange terms. Accordingly, the equations for the transverse magnetizations are:

$$\frac{\partial m^{(A)}}{\partial t} = -i2\pi\gamma B_z^{(Exp)} m^{(A)} - \frac{m^{(A)}}{T_2^{(A)}} + i2\pi\gamma B_{1T}^{(Exp)} M_z^{(A)}$$

$$+ \left(\frac{\partial m^{(A)}}{\partial t}\right)_{(kin)} + \left(\frac{\partial m^{(A)}}{\partial t}\right)_{(ex)} \qquad \text{(Eq. C3.7–2)}$$

and

$$\frac{\partial m^{(B)}}{\partial t} = -i2\pi\gamma B_z^{(Exp)} m^{(B)} - \frac{m^{(B)}}{T_2^{(B)}} + i2\pi\gamma B_{1T}^{(Exp)} M_z^{(B)} + \left(\frac{\partial m^{(B)}}{\partial t}\right)_{(ex)}$$

$$\text{(Eq. C3.7–3)}$$

and for the longitudinal magnetizations:

$$\frac{\partial M_z^{(A)}}{\partial t} = -2\pi\gamma \text{Im}\{B_{1T}^{(Exp)*}m^{(A)}\} + \frac{(M_z^{(A\_eq)} - M_z^{(A)})}{T_1^{(A)}}$$

$$+ \left(\frac{\partial M_z^{(A)}}{\partial t}\right)_{(kin)} + \left(\frac{\partial M_z^{(A)}}{\partial t}\right)_{(ex)} \qquad \text{(Eq. C3.7–4)}$$

and

$$\frac{\partial M_z^{(B)}}{\partial t} = -2\pi\gamma \text{Im}\{B_{1T}^{(Exp)*}m^{(B)}\} + \frac{(M_z^{(B\_eq)} - M_z^{(B)})}{T_1^{(B)}} + \left(\frac{\partial M_z^{(B)}}{\partial t}\right)_{(ex)}$$

$$\text{(Eq. C3.7–5)}$$

As discussed in the preceding section, the kinetic terms of the liquid pool and are given by:

$$\left(\frac{\partial m^{(A)}}{\partial t}\right)_{(kin)} \equiv \vec{\nabla}\cdot\mathbf{D}^{(A)}\cdot\vec{\nabla}m^{(A)} - \vec{v}^{(A)}\cdot\vec{\nabla}m^{(A)} \qquad \text{(Eq. C3.7–6)}$$

and

$$\left(\frac{\partial M_z^{(A)}}{\partial t}\right)_{(kin)} \equiv \vec{\nabla}\cdot\mathbf{D}^{(A)}\cdot\vec{\nabla}(M_z^{(A)} - M_z^{(A\_eq)}) - \vec{v}^{(A)}\cdot\vec{\nabla}(M_z^{(A)} - M_z^{(A\_eq)})$$

$$\text{(Eq. C3.7–7)}$$

The corresponding kinetic terms for the semisolid pool have been purposely omitted under the assumption that the contributions to magnetization dynamics are negligibly small.

   Up to this point, we have not assumed particular forms for the magnetization-exchange time derivatives and these must be specified further to fully characterize the mathematical problem. Mathematically tractable equations result from the following five assumptions:

1) The exchange rate between the liquid and semisolid pools is characterized by a fundamental decay rate R, which is independent of the pool sizes.

2) The instantaneous net exchange of longitudinal magnetization between the two pools is null:

$$\left(\frac{\partial M_z^{(A)}}{\partial t}\right)_{(ex)} + \left(\frac{\partial M_z^{(B)}}{\partial t}\right)_{(ex)} = 0 \qquad \text{(Eq. C3.7–8)}$$

3) The exchange of longitudinal magnetization between pools take place at constant rates and because the two pools are in steady state, the rate constants are the fundamental decay rate (R) multiplied by the population of the destination pool, specifically: $RM_z^{(B\_eq)}$ : A → B and $RM_z^{(A\_eq)}$ : B → A. Accordingly, the exchange terms of the general two-pool Bloch equations presented in the previous section, are:

$$\left(\frac{\partial M_z^{(A)}}{\partial t}\right)_{(ex)} = -\left(\frac{\partial M_z^{(B)}}{\partial t}\right)_{(ex)} = -RM_z^{(B\_eq)}M_z^{(A)} + RM_z^{(A\_eq)}M_z^{(B)}$$

$$\text{(Eq. C3.7–9)}$$

4) The magnetization exchange terms for the transverse magnetizations are assumed negligible

$$\left|\left(\frac{\partial m^{(A,B)}}{\partial t}\right)_{(ex)}\right| \ll \left|\frac{m^{(A,B)}}{T_2^{(A,B)}}\right| \qquad \text{(Eq. C3.7–10)}$$

5) Kinetic effects on longitudinal magnetization of the liquid pool are also assumed negligible.

$$\left(\frac{\partial M_z^{(A,B)}}{\partial t}\right)_{(kin)} \cong 0 \quad \text{and} \quad \left(\frac{\partial m^{(B)}}{\partial t}\right)_{(kin)} \cong 0$$

$$\text{(Eq. C3.7–11)}$$

Altogether, these five assumptions lead to a well-posed mathematical problem if all magnetic fields acting on the spin packet are known.

First for the transverse magnetization of the liquid and semisolid pools:

$$\frac{\partial m^{(A)}}{\partial t} = -i2\pi\gamma B_z^{(Exp)}m^{(A)} - \frac{m^{(A)}}{T_2^{(A)}} + i2\pi\gamma B_{1T}^{(Exp)}M_z^{(A)}$$

$$+\vec{\nabla}\cdot D^{(A)}\cdot\vec{\nabla}_m^{(A)} - \vec{v}^{(A)}\cdot\vec{\nabla}_m^{(A)} \qquad \text{(Eq. C3.7–12)}$$

and,

$$\frac{\partial m^{(B)}}{\partial t} = -i2\pi\gamma B_z^{(Exp)} m^{(B)} - \frac{m^{(B)}}{T_2^{(B)}} + i2\pi\gamma B_{1T}^{(Exp)} M_z^{(B)}$$

(Eq. C3.7–13)

Second, for the longitudinal magnetizations of the two pools:

$$\frac{\partial M_z^{(A)}}{\partial t} = -2\pi\gamma Im\{B_{1T}^{(Exp)*} m^{(A)}\} + \frac{(M_z^{(A\_eq)} - M_z^{(A)})}{T_1^{(A)}}$$

$$-RM_z^{(B\_eq)} M_z^{(A)} + RM_z^{(A\_eq)} M_z^{(B)}$$

(Eq. C3.7–14)

and,

$$\frac{\partial M_z^{(B)}}{\partial t} = -2\pi\gamma Im\{B_{1T}^{(Exp)*} m^{(B)}\} + \frac{(M_z^{(B\_eq)} - M_z^{(B)})}{T_1^{(B)}}$$

$$+RM_z^{(B\_eq)} M_z^{(A)} - RM_z^{(A\_eq)} M_z^{(B)}$$

(Eq. C3.7–15)

In practice however, modified versions of these equations are used. In particular, in the two-pool model (Henkelman *et al.*, 1993), which is also known as the binary spin bath model, the kinetic effects of the liquid pool are neglected and therefore the equations for the two complex transverse magnetizations are of identical form, specifically:

$$\frac{\partial m^{(A,B)}}{\partial t} = i2\pi\Delta v m^{(A,B)} - \frac{m^{(A,B)}}{T_2^{(A,B)}} - i\omega_1(t) M_z^{(A,B)}$$

(Eq. C3.7–16)

These have been written in a frame of reference rotating about the z-axis at the Larmor frequency; this is known as the rotating frame. Furthermore, the equations for the longitudinal magnetizations are modified to account for the fact that the semisolid pool has a different absorption lineshape (Morrison, Stanisz, & Henkelman, 1995), which deviates from the usual Lorentzian lineshape of the liquid pool (see section E3.7). The two-pool model longitudinal Bloch equations in the rotating frame are (Henkelman, *et al.*, 1993):

$$\frac{\partial M_z^{(A)}}{\partial t} = \frac{(M_z^{(A\_eq)} - M_z^{(A)})}{T_1^{(A)}} - RM_z^{(B\_eq)} M_z^{(A)}$$

$$+RM_z^{(A\_eq)} M_z^{(B)} + \omega_1(t) M_y^{(A)}$$

(Eq. C3.7–17)

and,

$$\frac{\partial M_z^{(B)}}{\partial t} = \frac{(M_z^{(B\text{-}eq)} - M_z^{(B)})}{T_1^{(B)}} - (R_{rfB}(\Delta v)$$
$$+ RM_z^{(A\text{-}eq)})M_z^{(B)} + RM_z^{B\text{-}eq}M_z^{(A)}$$

$$(\text{Eq. C3.7–18})$$

where $R_{rfb}(\Delta v)$ is the rate of loss of longitudinal magnetization by the semisolid pool due to off-resonance irradiation of amplitude $\omega_1(t) = 2\pi\gamma B_1(t)$ and offset frequency $\Delta v$.

## C3.8. *Relaxation by Magnetization Transfer*

During the application of multislice MRI pulse sequences, the semisolid pool of a given slice can be driven to equilibrium by the rf pulses acting on other slices (Dixon, Engels, Castillo, & Sardashti, 1990; Melki & Mulkern, 1992; G. Santyr, 1993). These rf pulses act as off resonance pulses and for rf intensive pulse sequences it is reasonable to assume that the semisolid pool is driven into a state of equilibrium, accordingly:

$$\frac{\partial M_z^{(B)}}{\partial t} \cong 0 \qquad (\text{Eq. C3.8–1})$$

Under this assumption, the BTS equation (Eq. C3.7–14) for the longitudinal magnetization of the liquid pool can be decoupled from the dynamic effects of the semisolid pool resulting in:

$$\frac{\partial M_z^{(A)}}{\partial t} = -2\pi\gamma \text{Im}\{B_{1T}^{(Exp)*}m^{(A)}\} + \left(\left[\frac{1}{T_1^{(A)}}\right]_{(kin)} + RM_z^{(B\text{-}eq)}\right)$$
$$\times (M_z^{(A\text{-}eq)} - M_z^{(A)}) + \text{Constant} \qquad (\text{Eq. C3.8–2})$$

this is the standard single pool longitudinal magnetization Bloch equation with a different total longitudinal magnetization relaxation rate, specifically

$$\frac{1}{T_1^{(A\text{-}obs)}} = \left[\frac{1}{T_1^{(A)}}\right]_{(kin)} + RM_z^{(B\text{-}eq)} \qquad (\text{Eq. C3.8–3})$$

This shows explicitly that when removed from equilibrium, the longitudinal magnetization of liquid pool still equilibrates monoexponentially but at a faster rate than in the absence of a semisolid pool. Furthermore, the difference in longitudinal relaxation rates is equal to the magnetization exchange rate $RM_z^{(B\text{-}eq)}$ : A → B, which in the two pool model is assumed constant. Hence, one of the most basic manifestations of magnetization exchange with a semisolid pool driven to equilibrium is faster longitudinal magnetization equilibration. Conversely, if the longitudinal magnetization of the semisolid pool is not constant, then the recovery of the liquid longitudinal magnetization may not be monoexponential.

### C3.9. *The Spin-Lock Bloch Equations*

The simplest and most basic Bloch equation problem is that of a non-relaxing spin packet subject to only the polarizing field. In this case, the experimental magnetic field is:

$$\vec{B}^{(\text{Exp})} = B_0\,\hat{z} \qquad \text{(Eq. C3.9–1)}$$

Furthermore, the Bloch equation without relaxation terms, specifically:

$$\frac{d\vec{M}}{dt} = 2\pi\gamma\vec{M} \times B_0\hat{z} \qquad \text{(Eq. C3.9–2)}$$

can be easily solved since:

$$\frac{dM_z}{dt} = 0 \qquad \text{(Eq. C3.9–3)}$$

and also,

$$\frac{d|\vec{M}|^2}{dt} = 0 \qquad \text{(Eq. C3.9–4)}$$

Moreover, the complex transverse magnetization satisfies:

$$\frac{dm}{dt} = -i2\pi\gamma B_0 m \qquad \text{(Eq. C3.9–5)}$$

which can be readily integrated leading to:

$$m(t) = m(0)\exp(-i2\pi\gamma B_0 t) \qquad \text{(Eq. C3.9–6)}$$

As such, the magnetization precesses about the z-axis with constant inclination angle as well as a constant amplitude thus defining cone of precession. Therefore, in the absence of relaxation, the magnetization will continually precess on the surface of a cone with a fixed inclination angle and it will never reach thermal equilibrium. If on the other hand the

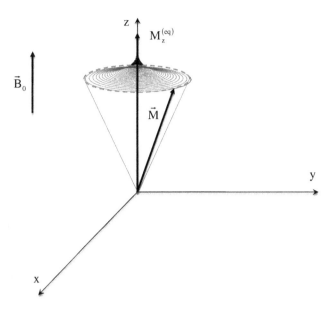

Fig. C-3. The magnetization dynamics of a non-relaxing spin packet in a constant polarizing field. The magnetization vector rotates on the surface of a cone at the Larmor frequency. In the absence of relaxation, such motion is perpetual. The effect of relaxation is also illustrated: the magnetization spirals inward from the cone of precession toward the z-axis.

relaxation mechanisms become active, then the magnetization would spiral down from the cone of precession towards the z-axis while its magnitude would increase to the longitudinal equilibrium value (see Fig. C-3).

We study now the effect of superposing an alternating magnetic field that rotates in the transverse plane with constant angular frequency $\omega_{rf}$, accordingly the experimental magnetic field in this case is:

$$\vec{B}^{(\text{Exp})}(t) = B_0\, \widehat{z} + B_1(t)(\cos(\omega_{rf}t)\hat{x} + \sin(\omega_{rf}t)\hat{y})$$

(Eq. C3.9–7)

The non-relaxing Bloch equation, specifically:

$$\frac{d\vec{M}}{dt} = 2\pi\gamma\vec{M} \times (B_0\, \widehat{z} + B_1(t)(\cos(\omega_{rf}t)\hat{x} + \sin(\omega_{rf}t)\hat{y}))$$

(Eq. C3.9–8)

can be recast in a frame of reference rotating at $\omega_{rf}$ about the z-axis and in the same direction as the applied $B_1$ field, leading to:

$$\frac{d\vec{M}^{(r)}}{dt} = 2\pi\gamma\vec{M}^{(r)} \times \vec{B}_{eff} \qquad \text{(Eq. C3.9–9)}$$

where the effective magnetic field (see Fig. C-4) is given by:

$$\vec{B}_{eff} = \left(B_0 - \frac{\omega_{rf}}{2\pi\gamma}\right)\hat{z}' + B_1(t)\hat{x}' = |\vec{B}_{eff}|\hat{\rho} \qquad \text{(Eq. C3.9–10)}$$

Hence, the Bloch equation in the rotating frame has the exact same form of the simpler previous example in the laboratory frame, and consequently, in the absence of relaxation the magnetization will precess about about a fixed axis, in this case in the direction of the effective magnetic field. The spatial coordinate for this new precession axis is universally denoted as $\rho$ and when relaxation effects are incorporated, this leads to relaxation being parametized with the relaxation times in the rotating frame, which are known as $T_{1\rho}$ and $T_{2\rho}$. In other words, in the rotating frame the magnetization is locked the the $\rho$-axis at an angle $\Phi$ of

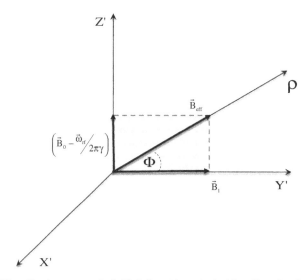

Fig. C-4.   The effective magnetic field defines the spin-locking direction in the rotating frame of reference.

the Y' axis (see Fig. C-4) given by:

$$\Phi = \arctan\left(\frac{(B_0 - \omega_{rf}/2\pi\gamma)}{B_1}\right) \qquad \text{(Eq. C3.9–11)}$$

Furthermore, the magnetization relaxes in the rotating frame towards an equilibrium value equal to (Moran & Hamilton, 1995):

$$M_\rho^{(eq)} = \frac{T_{1\rho}}{T_1} M_z^{(eq)} \sin(\Phi) \qquad \text{(Eq. C3.9–12)}$$

with the following relaxation times; for the longitudinal component:

$$\frac{1}{T_{1\rho}} = \frac{1}{T_1}\left[\sin^2(\Phi) + \frac{T_1}{T_2}\cos^2(\Phi)\right] \qquad \text{(Eq. C3.9–13)}$$

and for the transverse component:

$$\frac{1}{T_{2\rho}} = \frac{1}{2}\left[\frac{1}{T_2} + \frac{1}{T_1}\cos^2(\Phi) + \frac{1}{T_2}\sin^2(\Phi)\right] \qquad \text{(Eq. C3.9–14)}$$

When $\omega_{rf}$ is chosen closer to the Larmor frequency, $\Phi$ is lowered and therefore:

$$\frac{1}{T_{1\rho}} \xrightarrow{\ \Phi \longrightarrow 0\ } \frac{1}{T_2} \qquad \text{(Eq. C3.9–15)}$$

and for the transverse component:

$$\frac{1}{T_{2\rho}} \xrightarrow{\ \Phi \longrightarrow 0\ } \frac{1}{2}\left(\frac{1}{T_1} + \frac{1}{T_2}\right) \qquad \text{(Eq. C3.9–16)}$$

The corresponding spin-lock Bloch equations (Moran & Hamilton, 1995; Taheri & Sood, 2006) are:

$$\frac{dM_\rho}{dt} = \frac{(M_\rho^{(eq)} - M_\rho)}{T_{1\rho}} - \frac{d\Phi}{dt}M_\Psi \qquad \text{(Eq. C3.9–17)}$$

and,

$$\frac{dM_\Psi}{dt} = -\frac{M_\Psi}{T_{2\rho}} - 2\pi\gamma B_{eff}M_x + \frac{d\Phi}{dt}M_\rho \qquad \text{(Eq. C3.9–18)}$$

and,

$$\frac{dM_x}{dt} = -\frac{M_x}{T_{2\rho}} + 2\pi\gamma B_{eff}M_\Psi \qquad \text{(Eq. C3.9–19)}$$

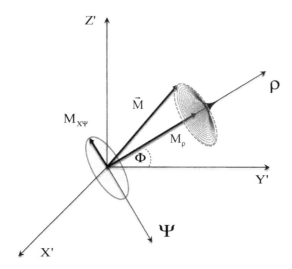

Fig. C-5.  The principle of spin-lock imaging.

If $d\Phi/dt \approx 0$, which can be accomplished by appropriately designing the
the $B_1$ envelope, then the first two spin-lock Bloch equations are decoupled
and spin-tip is avoided. In this case, the magnetization relaxes toward
equilibrium in the simple spiral form as illustrated in Fig. C-5 and the
longitudinal component along the spin-lock direction satisfy the simple
equation:

$$\frac{dM_\rho}{dt} = \frac{(M_\rho^{eq} - M_\rho)}{T_{1\rho}} \qquad \text{(Eq. C3.9–20)}$$

which predicts exponential recovery with a relaxation time $T_{1\rho}$.

To the extent that the effective magnetic field can be made arbitrarely
small, the study of $T_{1\rho}$ becomes of increasing interest as it provides insight
the slow molecular motions in tissue.

### C3.10.  *The Semi-Classical Classification Scheme*
###         *of qNMR-Parameters*

Altogether, the two-pool BTS differential equations (Eq. C3.7-12-15) of
the preceding section fully characterize two-pool spin packets provided
that: 1) all qNMR parameters and 2) all magnetic fields are known.
Furthermore, the equilibrium longitudinal magnetizations are proportional

to the corresponding spin densities, specifically:

$$M_z^{(A\text{-}eq)} = \left(\frac{\pi^2\gamma^2\hbar^2 B_0}{k_B T}\right) PD^{(A)} \qquad \text{(Eq. C3.10-1)}$$

and,

$$M_z^{(B\text{-}eq)} = \left(\frac{\pi^2\gamma^2\hbar^2 B_0}{k_B T}\right) PD^{(B)} \qquad \text{(Eq. C3.10-2)}$$

The phenomenological qNMR parameters of the BTS equations report on the states of: 1) $^1$H-proton pools, in terms of thermal equilibrium spin density, 2) kinetics, and 3) interactions with the microenvironment *via* the relaxation times. Accordingly, the qNMR parameters of the liquid pool can be organized in tabular form as follows:

$$\text{qNMR\_par}^{(A)} = \left\{ \begin{matrix} PD^{(A)} & \cdots & \cdots & \cdots \\ \vdots & \mathbf{D}^{(A)} & \vec{v}^{(A)} & \\ \vdots & T_1^{(A)} & T_2^{(A)} & T_{1,2\rho}^{(A)} \end{matrix} \right\}$$

$$\text{(Eq. C3.10-3)}$$

where the first, second, and third rows consist of spin density, kinetic, and interaction parameters respectively. Analogously for the semisolid pool, which is coupled to the liquid pool *via* the fundamental decay rate (R), we write:

$$\text{qNMR\_par}^{(B)} = \left\{ \begin{matrix} PD^{(B)} & \cdots & \cdots & \cdots \\ \vdots & \mathbf{D}^{(B)} \cong 0 & \vec{v}^{(B)} \cong 0 & \\ \vdots & T_1^{(B)} & T_2^{(B)} & \end{matrix} \right\}$$

$$\text{(Eq. C3.10-4)}$$

where we have assumed that the semisolid pool is rigid and static.

The two-pool NMR model of spin packets is complete when the totality of qNMR parameters of both spin pools plus the inter-pool coupling constant R is known. Considering that the diffusion tensor and velocity consist of 6 and 3 independent parameters respectively, the total number of qNMR parameters amount to 14 parameters for the liquid pool, 3 parameters for the semisolid pool, and one inter-pool coupling parameter, altogether,

$$\text{qNMR\_par}^{(A)}\{14\} \overset{R\{1\}}{\longleftrightarrow} \text{qNMR\_par}^{(B)}\{3\} \qquad \text{(Eq. C3.10–5)}$$

These 18 primary qNMR parameters can be further processed for deriving secondary qNMR parameters, which may be more practical and/or useful for scientific and medical purposes. There are innumerable ways of combining the primary parameters; important examples include secular-$T_2$, the longitudinal-to-transverse relaxation time ratio, the eigenvalues and eigenvectors of the diffusion tensor, the bound water fraction, and many others.

We note that the concept of tissue perfusion is undefined at the spatial scale of spin packets because tissue microvasculature has a characteristic spatial scale that is several orders of magnitude larger than spin packets and therefore perfusion parameters become meaningful at the imaging voxel scale, as will be discussed later. Analogously, the concept of $T_2^*$ is undefined or unnecessary at the spin packet scale.

Solving the two-pool BTS differential equations also requires knowing *a priori* the actual experimental magnetic field for all positions and times, specifically

$$\vec{B}^{(\text{Exp})}(\vec{x}, t) = \vec{B}^{(\text{Theo})}(\vec{x}, t) + \delta\vec{B}_0(\vec{x}) + \delta\vec{B}_1^{(\text{rf})}(\vec{x}, t) + \delta\vec{B}^{(\text{enc})}(\vec{x}, t)$$

$$\text{(Eq. C3.10–6)}$$

As discussed earlier in section C1.4, only the theoretically intended magnetic field is known beforehand, specifically

$$\vec{B}^{(\text{Theo})}(\vec{x}, t) = B_0\,\hat{z} + B_1(t)(\cos(\omega_{\text{rf}}t)\hat{x} + \sin(\omega_{\text{rf}}t)\hat{y} + \vec{x}\cdot\vec{g}(t)\hat{z}$$

$$\text{(Eq. C3.10–7)}$$

and the magnetic field deviations must be determined experimentally; these are typically measured or estimated by methods that are conceptually similar to those used for measuring the qNMR parameters of Equations C3.10-3, 4.

## C4. Essential NMR Dynamics

The Bloch-Torrey equations describe the fundamental aspects of magnetization dynamics, specifically: 1) the ability of manipulating the magnetization *via* nuclear magnetic resonance (excitation and refocusing), and 2) the processes of thermal equilibration *via* $T_1$ and $T_2$ relaxation. This section is concerned with solving five classic Bloch equation problems that illustrate the main physical processes of NMR dynamics, specifically the processes of: 1) NMR excitation, 2) inversion recovery of longitudinal magnetization, 3) magnetization saturation, 4) free induction decay, and 5) spin-echoes. For simplicity, this section is concerned only with single-pool spin packets --*i.e.* the liquid pool-- and the superscript (A) is assumed and omitted for this section only.

## C4.1. *NMR Excitation: Flip Angle and Slice Selection*

We study the liquid pool Bloch equations for spin packets that are magnetized by a perfectly homogeneous polarizing field and in addition, with a pulsed field gradient superimposed. All spin packets are assumed to be in thermal equilibrium at time zero and may have different proton densities depending on position, thus leading to position-dependent initial conditions:

$$m(\vec{x}, 0) = 0 \quad \text{and} \quad M_z(\vec{x}, 0) = M_z^{(eq)}(\vec{x}) \qquad \text{(Eq. C4.1–1)}$$

We further consider the dynamic effects of applying a circularly polarized active $B_1$ field pulse. Accordingly, the full experimental magnetic field is given by:

$$\vec{B}^{Exp}(t) = (B_0 + \vec{x} \cdot \vec{g}(t))\hat{z} + B_1(t)(\cos(\omega_{rf}t + \varphi_{rf})\hat{x} + \sin(\omega_{rf}t + \varphi_{rf})\hat{y})$$
$$\text{(Eq. C4.1–2)}$$

The transverse $B_1$ phasor reads:

$$B_{1T}^{(Exp)} = B_1(t) \exp[i(\omega_{rf}t + \varphi_{rf})] \qquad \text{(Eq. C4.1–3)}$$

We will prove that under conditions of magnetic resonance, the magnetization vector can be forcedly rotated by any specified flip angle around any desired axis contained in the transverse plane.

We study the magnetization dynamics with short excitatory slice selective pulses: for times $t \ll T_2, T_1$ the relaxation terms of the transverse Bloch equation can be neglected leading to:

$$\frac{\partial m}{\partial t} = -i2\pi\gamma(B_0 + \vec{x} \cdot \vec{g}(t))m + i2\pi\gamma B_1(t) \exp[i(\omega_{rf}t + \varphi_{rf})]M_z$$
$$\text{(Eq. C4.1–4)}$$

By writing the transverse magnetization in terms of a modulus and a phase factor, specifically

$$m(\vec{x}, t) = |m(\vec{x}, t)| \exp(i\Phi(\vec{x}, t)) \qquad \text{(Eq. C4.1–5)}$$

We derive the following system of differential equations for the magnitude and phase of the transverse magnetization:

$$\frac{\partial |m|}{\partial t} = 2\pi\gamma B_1(t) M_z \sin(\Phi - \omega_{rf} t - \varphi_{rf}) \qquad \text{(Eq. C4.1–6)}$$

and

$$\frac{\partial \Phi}{\partial t} = -2\pi\gamma (B_0 + \vec{x} \cdot \vec{g}(t)) + 2\pi\gamma B_1(t) \left(\frac{M_z}{|m|}\right) \cos(\Phi - \omega_{rf} t - \varphi_{rf})$$

$$\text{(Eq. C4.1–7)}$$

Additionally, for times $t \ll T_2, T_1$ the longitudinal Bloch equation for the z-component reads:

$$\frac{\partial M_z}{\partial t} = -2\pi\gamma B_1(t)|m| \sin(\Phi - \omega_{rf} t - \varphi_{rf}) \qquad \text{(Eq. C4.1–8)}$$

It can be easily shown that these differential equations predict that the modulus of the magnetization vector is constant in time and that therefore:

$$|m|^2 + M_z^2 = M_z^{(eq)2} \qquad \text{(Eq. C4.1–9)}$$

Using the equality above, the longitudinal Bloch equation can be rewritten as:

$$\frac{\partial M_z}{\partial t} = 2\pi\gamma B_1(t) \sqrt{M_z^{(eq)2} - M_z^2} \sin(\Phi - \omega_{rf} t - \varphi_{rf})$$

$$\text{(Eq. C4.1–10)}$$

This last equation can be readily integrated leading to the following exact integral solution:

$$M_z(\vec{x}, t) - i|m(\vec{x}, t)| = M_z^{(eq)} \exp\left[-i2\pi\gamma \int_0^t B_1(t')\right.$$

$$\left. \times \sin(\Phi(\vec{x}, t') - \omega_{rf} t' - \varphi_{rf}) dt'\right]$$

$$\text{(Eq. C4.1–11)}$$

Furthermore, by identifying the real and imaginary parts of both sides of the equation above, we find:

$$|m(\vec{x}, t)| = M_z^{(eq)} \sin\left[2\pi\gamma \int_0^t B_1(t') \sin(\Phi(\vec{x}, t') - \omega_{rf}t' - \varphi_{rf})dt'\right]$$

(Eq. C4.1–12)

and

$$M_z(\vec{x}, t) = M_z^{(eq)} \cos\left[2\pi\gamma \int_0^t B_1(t') \sin(\Phi(\vec{x}, t') - \omega_{rf}t' - \varphi_{rf})dt'\right]$$

(Eq. C4.1–13)

With these last two equations, we can calculate the instantaneous flip angle --also known as, tip angle-- is defined in terms of the transverse to longitudinal magnetization ratio *via:*

$$tg(FA) = \frac{|m|}{M_z}$$

(Eq. C4.1–14)

Therefore, we find an exact formula for the flip angle, specifically:

$$FA(\vec{x}, t) = 2\pi\gamma \int_0^t B_1(t') \sin[\Phi(\vec{x}, t') - \omega_{rf}t' - \varphi_{rf}]dt'$$

(Eq. C4.1–15)

which is a function of the instantaneous phase of the transverse magnetization, given by:

$$\Phi(\vec{x}, t) = \Phi(0) - 2\pi\gamma \int_0^t \{B_0 + \vec{x} \cdot \vec{g}(t') + B_1(t')$$

$$\times \cot(FA) \cos(\Phi - \omega_{rf}t' - \varphi_{rf})\}dt'$$

(Eq. C4.1–16)

This last result is obtained by integrating Eq. C4.1–7 with respect to time. In summary, the Bloch equation problem of slice selective excitation without relaxation terms, leads to a set of two nonlinear integral equations that are coupled.

### C4.1.i. *The Resonance Condition*

Notably, the system of equations above for the flip angle and the transverse magnetization phase decouples if the carrier rf frequency satisfies:

$$\sin(\Phi(\vec{x}, t) - \omega_{rf} t - \varphi_{rf}) = 1 \qquad \text{(Eq. C4.1–17)}$$

which as we will show, leads to the condition of (exact) nuclear magnetic resonance.

With this assumption, the flip angle resulting from applying the $B_1$ pulse is given by the well known on resonance flip angle formula:

$$FA(t) = 2\pi\gamma \int_0^t B_1(t') dt' \qquad \text{(Eq. C4.1–18)}$$

which is position independent. Furthermore, the phase of the transverse magnetization is given by:

$$\Phi(\vec{x}, t) = \omega_{rf} t + \varphi_{rf} + \frac{\pi}{2} \qquad \text{(Eq. C4.1–19)}$$

and also by:

$$\Phi(\vec{x}, t) = \Phi(0) - 2\pi\gamma \int_0^t \{B_0 + \vec{x} \cdot \vec{g}(t')\} dt' \qquad \text{(Eq. C4.1–20)}$$

In the absence of an applied gradient ($\vec{g}(t) = 0$), these last two equations can be satisfied simultaneously, leading to:

$$\omega_{rf} = -\omega_0 \quad \text{and} \quad \Phi(0) = \varphi_{rf} + \frac{\pi}{2} \qquad \text{(Eq. C4.1–21)}$$

Meaning that for a $B_1$ pulse with central carrier frequency equal to the Larmor frequency, efficient conversion of longitudinal into transverse magnetization takes place, and the newly generated transverse magnetization is orthogonal to $B_1$ at all times. Such $B_1$ magnetic field rotates in the same direction as the precessing magnetization (see Eq. C2.2–30); clockwise as seen from above.

$$B_{1T}^{(Exp)} = B_1(t) \exp[-i(\omega_0 t - \varphi_{rf})] \qquad \text{(Eq. C4.1–22)}$$

In particular, if $\varphi_{rf} = 0$, the applied $B_1$ field in the rotating frame is along the x-axis, which is the rotation axis; accordingly, the flip angle is designated accordingly as $FA_x$. The net effect of applying such an rf pulse can be expressed by:

$$M_z^{(eq)}(\vec{x}) \xrightarrow{\ FA_x\ } m(\vec{x}, 0) = iM_z^{(eq)}(\vec{x}) \sin(FA_x) \quad \text{(Eq. C4.1–23)}$$

In summary, in the absence of applied magnetic field gradients ($\vec{g}(t) = 0$), the Bloch equation problem of NMR excitation can be solved analytically. More specifically, the equations for the flip and for the transverse magnetization phase can be decoupled and solved in close form leading to the condition of exact resonance.

## C4.1.ii. *Slice Selection*

We study now the effect of applying simultaneously an rf field and a constant magnetic field gradient, *i.e.* $\vec{g}(t) = \vec{g}_{ss} \neq 0$, where we have used the subscript "ss" to denote slice selection. As we will show in this section, the net effect of this combination is to excite only tissues within a slice that is orthogonal to the vector $\vec{g}_{ss}$.

In this case, the flip angle equation and transverse magnetization phase integral equation, specifically:

$$FA(\vec{x}, t) = 2\pi\gamma \int_0^t B_1(t') \sin[\Phi(\vec{x}, t') - \omega_{rf}t' - \varphi_{rf}]dt'$$

$$\text{(Eq. C4.1–24)}$$

and

$$\Phi(\vec{x}, t) = \Phi(0) - 2\pi\gamma \int_0^t \{B_0 + \vec{x} \cdot \vec{g}(t') + B_1(t') \cot(FA)$$

$$\times \cos(\Phi - \omega_{rf}t' - \varphi_{rf})\}dt' \quad \text{(Eq. C4.1–25)}$$

are not easily decoupled. We resort to integrating these equations approximately using the fact that the last term in Eq. C4.1-24 is negligible near resonance: $\cos(\Phi - \omega_{rf}t' - \varphi_{rf}) \approx 0$, hence:

$$\Phi(\vec{x}, t) \cong \Phi(0) + \omega_0 t - 2\pi\gamma\vec{x} \cdot \vec{g}_{ss}t \quad \text{(Eq. C4.1–26)}$$

In addition, we express the rf field envelope in terms of its Fourier transform, specifically:

$$B_1(t) = \int_{-\infty}^{\infty} \hat{B}_1(\omega) \exp[i\omega t]d\omega \quad \text{(Eq. C4.1–27)}$$

By substituting these last two expressions in the integral equation for the flip angle and by interchanging the order of integration, we find:

$$FA(\vec{x}, T) = 2\pi\gamma \int_{-\infty}^{\infty} \hat{B}_1(\omega) \left\{ \int_0^T \exp[i\omega t'] \sin[(\omega_0 - \omega_{rf}) \right.$$

$$\left. -2\pi\gamma\vec{x} \cdot \vec{g}_{ss}t' + \Phi(0) - \varphi_{rf}]dt' \right\} d\omega \qquad \text{(Eq. C4.1–28)}$$

where T is the duration of the rf pulse. The temporal integral inside the curly brackets is a function of frequency that is significantly different from zero only for frequencies near $\omega(\vec{x})$ given by:

$$\omega(\vec{x}) = -2\pi\gamma\vec{x} \cdot \vec{g}_{ss} \qquad \text{(Eq. C4.1–29)}$$

For this, we have chosen the carrier frequency of the rf pulse equal to the Larmor frequency.

Hence, to a first approximation the flip angle is position dependent and is given by:

$$\boxed{FA(\vec{x}, T) = 2\pi\gamma\hat{B}_1(\omega(\vec{x})) \qquad \text{(Eq. C4.1–30)}}$$

where $\hat{B}_1(\omega)$ is the Fourier transform of $B_1(t)$.

In conclusion, the profile of flip angles across the slice follows the profile of the excitatory rf pulse in Fourier domain. This justifies the use of rf pulses with sinc $(= \sin(x)/x)$ envelopes in time domain, which therefore result in constant flip angle through the slice. Hence, for a pulse of finite bandwidth, excitation occurs only inside a slice the width of which is determined by the bandwidth of the rf pulse.

It has been shown however that the flip angle equation above is not accurate in the limit of large flip angles (Joseph, Axel, & O'Donnell, 1984) and that the profile of a slice selective inversion pulse based on a sinc envelope can deviate significantly from the nominally intended 180° flip angle. A solution to this problem has been given by using a complex hyperbolic secant (sech) pulse (Silver, Joseph, & Hoult, 1985). This pulse belongs to class of waveforms based on analytical solutions of the Bloch equation (Rosenfeld, Panfil, & Zur, 1997). The mathematical solution indicates that under the appropriate conditions, the use of such a complex sech pulse creates a highly selective population inversion, which, above a critical threshold, is independent of pulse amplitude and hence independent of $B_1$ field homogeneity (Silver, *et al.*, 1985).

## C4.2. *Inversion Recovery*

We study the temporal evolution of the longitudinal magnetization of a spin packet that was removed from the state of equilibrium at time zero by a short $B_1$ pulse with a flip angle (FA). We are interested in solving the longitudinal Bloch equation in the absence of applied rf fields, accordingly:

$$\frac{\partial M_z}{\partial t} = \frac{(M_z^{(eq)} - M_z)}{T_1} \qquad \text{(Eq. C4.2–1)}$$

Furthermore, we will use the following initial condition: $M_z(0) = M_z^{(eq)}(\vec{x})\cos(FA_x)$. Hence:

$$M_z(t) = M_z^{(eq)}\left(1 - (1 - \cos(FA_x))\exp\left(-\frac{t}{T_1}\right)\right) \qquad \text{(Eq. C4.2–2)}$$

This solution has a null timepoint given by:

$$t_{null} = \ln(1 - \cos(FA_x))T_1 \qquad \text{(Eq. C4.2–3)}$$

For the special case of a full inversion pulse ($M_z(0) = -M_z^{(eq)}$), we find the well-known inversion recovery result:

$$M_z(t) = M_z^{(eq)}\left(1 - 2\exp\left(\frac{t}{T_1}\right)\right) \qquad \text{(Eq. C4.2–4)}$$

which describes an exponential recovery process that starts with the magnetization fully inverted along the negative z-axis, then passes through a null point at $t_{null} = \ln(2)T_1$, and finally continues to recover exponentially towards the equilibrium value along the positive z-axis.

## C4.3. *Magnetization Saturation in Repetitive Pulse Sequences*

MRI pulse sequences consist of several rf pulses that are played out repetitively in time intertwined with gradient pulses in the so-called imaging cycle or TR cycle. If the time of repetition (TR) of the pulse sequence is short relative to the time needed by the sample to fully relax $TR < 5\,T_1$, the longitudinal magnetization will not fully recover from period to period thus leading to magnetization saturation. In the steady state, the phenomenon of magnetization saturation consists in the reduction of longitudinal magnetization available to create transverse magnetization in each pulse sequence cycle, and therefore leads to a reduction in the MR signal intensity. During an imaging experiment, because different voxels

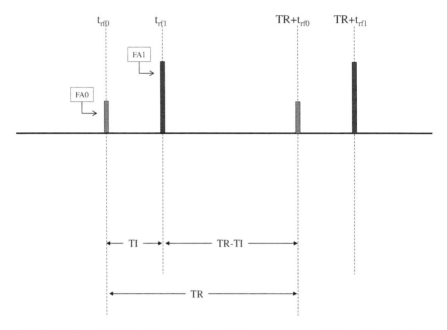

Fig. C-6. General timing diagram of a repetitive sequence consisting of two rf pulses.

will contain tissues of different $T_1$ values, differential voxel saturation leads to image contrast by $T_1$-weighting.

We analyze a repetitive series of two rf pulses (see **Fig.** C-6). We assume that each rf pulse rotates the magnetization by the x-axis. The net effect of applying an rf pulse is $M_z(t_{rf}^{(+)}) = \cos(FA)M_z(t_{rf}^{(-)})$; this assumes that the transverse magnetization is null at the time of each rf pulse. Between rf pulses, the longitudinal magnetization evolves according to the $T_1$-recovery equation:

$$M_z(t) = M_z^{(eq)} - (M_z^{(eq)} - \cos(FA)M_z(t_{rf}^{(-)}))E_1(t - t_{rf}^{(+)})$$

$$(Eq.\ C4.3\text{--}1)$$

where we have used the abbreviated exponential notation, specifically:

$$E_1(t - t_{rf}^{(+)}) \equiv \exp\left(-\frac{t - t_{rf}^{(+)}}{T_1}\right) \qquad (Eq.\ C4.3\text{--}2)$$

Basic methodology: We start the calculation at the time of the RF1 pulse and finish it a time TR after at which the following rf1 pulse is applied.

Then the steady state condition is applied, *i.e.*

$$M_z(t_{rf0,1}^{(-)}) = M_z(t_{rf0,1}^{(-)} + TR) = M_z^{sat\_0,1} \qquad \text{(Eq. C4.3--3)}$$

and we assumed that the transverse magnetization has been fully dephased form cycle to cycle. From which we calculate the partially saturated longitudinal magnetization at the time of rf1 pulse.

$$M_z^{sat\_1} = M_z^{(eq)} \left( \frac{1 - E_1(TI) + \cos(FA0)(E_1(TI) - E_1(TR))}{1 - \cos(FA0)\cos(FA1)E_1(TR)} \right)$$

$$\text{(Eq. C4.3--4)}$$

Analogously, the partially saturated longitudinal magnetization at the time of the RF0 pulse is

$$M_z^{sat\_0} = M_z^{(eq)} \left\{ 1 - (1 - \cos(FA1)) \right.$$
$$\left. \times \left( \frac{1 - E_1(TI) + \cos(FA0)(E_1(TI) - E_1(TR))}{1 - \cos(FA0)\cos(FA1)E_1(TR)} \right) \right) E_1(TR\_TI) \right\}$$

$$\text{(Eq. C4.3--5)}$$

These formulas are applied next to study saturation effects with three important pulse sequences, specifically gradient-echo, spin-echo, and inversion recovery.

1) **Gradient-echo** pulse sequence: for a train of identical pulses spaced by TR, the solution is:

$$M_z^{sat} = M_z^{(eq)} \frac{1 - E_1(TR)}{1 - \cos(FA)E_1(TR)} \qquad \text{(Eq. C4.3--6)}$$

and the transverse magnetization immediately after each rf pulse is:

$$|m^{sat}| = M_z^{(eq)} \sin(FA) \frac{1 - E_1(TR)}{1 - \cos(FA)E_1(TR)} \qquad \text{(Eq. C4.3--7)}$$

By differentiating with respect to the flip angle, we find the optimum flip angle for any given $T_1$, specifically

$$FA^{(opt)} = \cos^{-1} \left[ \exp\left( -\frac{TR}{T_1} \right) \right] \qquad \text{(Eq. C4.3--8)}$$

Such optimum flip angle is known as the Ernst angle and the equation above is often referred to as the Ernst equation (Ernst, Bodenhausen, & Wokaun, 1990). It gives the flip angle for which maximum signal is obtained from a tissue with a given $T_1$ being scanned at TR.

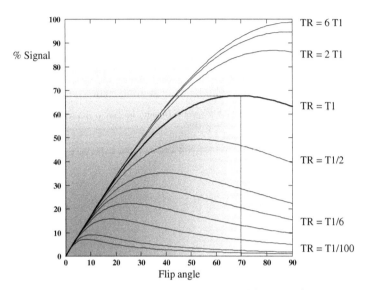

Fig. C-7. Magnetization saturation curves as function of flip angle (see Eq. C4.3–7) for various $TR/T_1$ ratios (gradient echo pulse sequences). The Ernst angle (see Eq. C4.3–8) is the flip angle at which maximum signal is obtained.

Magnetization saturation refers to the signal loss of a tissue, which is caused by exciting the tissue at a faster rate than it can recover according to its $T_1$ value. The graph shows that when TR is shorter than $T_1$, maximum SNR is obtained at a lower flip angle. It also shows the drastic overall SNR penalties, which can result from operating, pulse sequences at TRs much shorter than the tissue $T_1$s.

2) **Spin-echo (SE)**: in this case, FA0 = 90° and FA1 = 180°, cos(FA0) = 0 and cos(FA1) = −1. Hence, Eq. C4.3–5 reduces to the well-known SE signal saturation equation:

$$M_Z^{sat\_0} = M_Z^{(eq)} \left\{ 1 - \left( 2\exp\left( -\frac{(TR - TE/2)}{T_1} \right) - \exp\left( -\frac{TR}{T_1} \right) \right) \right\}$$

(Eq. C4.3–9)

where we have used $TI = TE/2$.

3) **Inversion recovery**: FA0 = 180° and FA1 = 90°, cos(FA0) = −1 and cos(FA1) = 0. In this case, Eq. C4.3–5 reduces to the well-known IR signal

saturation equation

$$M_z^{(sat\text{-}1)} = M_z^{(eq)} \left( 1 - 2\exp\left( -\frac{TI}{T_1} \right) + \exp\left( -\frac{TR}{T_1} \right) \right)$$

(Eq. C4.3–10)

## C4.4. Free Decay in the Presence of Magnetic Field Gradients: Effects of Diffusion

We consider a spin packet that was excited to the transverse state at time zero and we study the subsequent NMR dynamics in the presence of a time dependent magnetic field gradient, hence:

$$B_z^{(App)}(\vec{x}, t) = B_0 + \vec{x}\cdot\vec{g}(t)$$

(Eq. C4.4–1)

and therefore, the transverse BT equation reads:

$$\frac{\partial m}{\partial t} = -i(\omega_0 + 2\pi\gamma\vec{x} \cdot \vec{g}(t))m - \frac{m}{T_2} + D\nabla^2 m$$

(Eq. C4.4–2)

This transverse BT equation can be integrated exactly by the method of Stejskal and Tanner (Stejskal & Tanner, 1965) and the solution is:

$$m(\vec{x}, t) = m(\vec{x}, 0) \exp[-i\omega_0 t] \exp\left( -\frac{t}{T_2} \right)$$

$$\times \exp\left( -(2\pi\gamma)^2 D \int_0^t \left| \int_0^{t''} \vec{g}(t')dt' \right|^2 dt'' \right)$$

$$\times \exp\left( -i2\pi\gamma\vec{x} \cdot \int_0^t \vec{g}(t)dt' \right)$$

(Eq. C4.4–3)

Furthermore, by defining the k-vector of the phase factor as:

$$\vec{k}(t) \equiv \gamma \int_0^t \vec{g}(t')dt'$$

(Eq. C4.4–4)

and the b-factor of the freely decaying transverse magnetization by

$$b^{(FID)}(t) \equiv (2\pi)^2 \int_0^t |\vec{k}(t')|^2 dt'$$

(Eq. C4.4–5)

we obtain the following more compact result:

$$m(\vec{x}, t) = m(\vec{x}, 0) \exp[-i\omega_0 t] \exp\left[-\left(\frac{t}{T_2} + Db^{(FID)}(t)\right)\right]$$

$$\times \exp[-i2\pi\gamma\vec{x} \cdot \vec{k}(t)] \qquad \text{(Eq. C4.4–6)}$$

As will be discussed later, the preferred method for spatially encoding NMR signals is by using pulsed magnetic field gradients.

NMR signals as generated with the nuclear induction method, can be modeled mathematically as integrals of the transverse magnetization multiplied by the sensitivity profile $(\Omega(\vec{x}))$ of the receiving coil, *i.e.*

$$S(t) = \iiint_{\left\{\substack{\text{Infinite} \\ \text{space}}\right\}} \Omega(\vec{x})m(\vec{x}, t)d^3x \qquad \text{(Eq. C4.4–7)}$$

These last two equations establish the theoretical connection between gradient-encoded NMR signals and Fourier transform MR imaging. Specifically, with suitable choices of encoding gradient pulses, NMR signals can be generated and acquired such that the measurements constitute the Fourier transform of the transverse magnetization. In precise terms, such MRI signals in the rotating frame of reference --*i.e.* $S^{(rot)}(\vec{k}(t)) = S(\vec{k}(t)) \exp[+i\omega_0 t]$-- constitute the Fourier co-domain of the transverse magnetization in the rotating frame at the time of measurement:

$$S^{(rot)}(\vec{k}(t)) = \iiint_{\left\{\substack{\text{Infinite} \\ \text{space}}\right\}} \Omega(\vec{x})m(\vec{x}, 0) \exp\left[-\left(\frac{t}{T_2} + Db^{(FID)}(t)\right)\right]$$

$$\times \exp[-i2\pi\gamma\vec{x} \cdot \vec{k}(t)]d^3x \qquad \text{(Eq. C4.4–8)}$$

In practice, MRI signals are measured in the rotating frame by electronically removing the carrier frequency, and therefore the superscript "rot" can be omitted with little risk of confusion.

We note that the signal will be maximized when the measurements include the origin of k-space, thus the refocusing condition to be satisfied by the applied gradients in all directions:

$$\vec{k}(te) = \gamma \int_0^{te} \vec{g}(t')dt' = 0 \qquad \text{(Eq. C4.4–9)}$$

This integral is referred to as the zero order moment of the gradient waveform and therefore the condition for maximum MRI signal is to use compensated gradient waveforms with null zero order moments.

A very important application of Eq. C4-6 is the decay of transverse magnetization in the presence of an applied pulsed gradient plus a constant background gradient $(\vec{g}(t) = \vec{g}^{(app)}(t) + \vec{g}_0)$. The constant gradient, commonly referred to as background gradient, is of fundamental importance to MRI because its signal-weakening effects are in general unavoidable. Background gradients are typically caused by magnetic susceptibility inhomogeneities within the sample, and/or by scanner imperfections. In these circumstances:

$$m(\vec{x}, t) = m(\vec{x}, 0) \exp[-i\omega_0 t] \exp\left[-\frac{t}{T_2}\right] \exp[-Db^{(FID)}(t)]$$

$$\times \exp[-i2\pi\gamma\vec{x}\cdot(\vec{k}^{(app)}(t) + \vec{g}_0 t)] \qquad \text{(Eq. C4.4–10)}$$

where

$$b^{(FID)}(t) \equiv (2\pi)^2 \int_0^t |\vec{k}^{(app)}(t') + \gamma\vec{g}_0 t|^2 dt' \qquad \text{(Eq. C4.4–11)}$$

Mathematically, Eq. C4-10 implies that the refocusing condition $\vec{k}^{(app)}(te) + \vec{g}_0 te = 0$ may not be achieved at any time during the decay, particularly in regions of strong background gradients, thus potentially resulting in regions devoid of signal.

We note that if the applied gradient is zero, the familiar result is obtained.

$$m(\vec{x}, t) = m(\vec{x}, 0) \exp[-i\omega_0 t] \exp\left[-\frac{t}{T_2}\right] \exp\left[-\frac{(\pi\gamma)^2 |\vec{g}_0|^2 Dt^3}{3}\right]$$

$$\times \exp[-i2\pi\gamma\vec{x}\cdot\vec{g}_0 t] \qquad \text{(Eq. C4.4–12)}$$

which gives explicit terms for natural exponential $T_2$-decay, diffusive decay, and deconstructive interference due to dephasing. This equation will be instrumental for defining the decay time constant $T_2^*$ of the tissues in a voxel, as opposed to the decay time constant $T_2$ of spin packets.

## C4.5. *rf-Refocused Decay: Spin-Echoes*

We showed in the previous section that the refocusing condition might not be achieved at any time for systems evolving in the presence of gradient with non-zero first order moment; such is the case of constant background gradients. Hence, free magnetization decay experiments are vulnerable to signal loss in areas of background gradients, which are experimentally unavoidable as inhomogeneities to $B_0$.

In 1950, Erwin Hahn (Hahn, 1950) studied the temporal evolution of transverse magnetization for a sequence of two rf pulses, specifically a 90° excitatory pulse followed by a 180° refocusing pulse applied a time $\tau$ later. He showed experimentally and proved theoretically, that the background gradient-caused dephasing could be reversed leading to the formation of an echo signal at time $\tau$ after the refocusing pulse. Such rf-refocused signals are known a Hahn spin-echoes, or simply spin-echoes.

The effect of a 180° pulse applied at time $\tau$ and with a phase angle $\varphi$ relative to the 90° pulse, is to turn $m(\vec{x}, \tau)$ into $m^*(\vec{x}, \tau) \exp(i(2\varphi - \pi))$.

For the spin-echo pulse sequence, *i.e.* 90° $\xrightarrow{\tau}$ 180° $\xrightarrow{\tau}$ spin echo, solving the BT equation (Eq. C4.4-2) leads to:

$$
m(\vec{x}, t) = m(\vec{x}, 0) \exp[-i\omega_0 t] \exp\left[-\frac{t}{T_2}\right] \exp[-Db^{(SE)}(t)]
$$

$$
\times \exp[-i2\pi\vec{x} \cdot (\vec{k}(t) - 2H(t - \tau)\vec{k}(\tau))] \qquad \text{(Eq. C4.5–1)}
$$

where the function H is the Heaviside unit step function and is given by:

$$
H(t - \tau) = \begin{cases} 0 & \text{for} \quad 0 < t < \tau \\ +1 & \text{for} \quad t > \tau \end{cases} \qquad \text{(Eq. C4.5–2)}
$$

and the spin-echo b-factor is given by:

$$
b^{(SE)}(t) \equiv (2\pi)^2 \left\{ \int_0^t |\vec{k}(t')|^2 dt' - 4\vec{k}(\tau) \cdot \int_0^t \vec{k}(t')dt' + 4|\vec{k}(\tau)|^2(t - \tau) \right\}
$$

$$
\text{(Eq. C4.5–3)}
$$

This solution predicts the rf-refocused signal, *via* a spin-echo refocusing condition that reads:

$$
\vec{k}(TE) - 2H(TE - \tau)\vec{k}(\tau) = 0 \qquad \text{(Eq. C4.5–4)}
$$

which implies there exist a time $TE = 2\tau$ --known as the spin-echo time-- such that the refocusing condition is satisfied both, for the constant background gradient, and the pulsed applied gradients provided that the applied gradients are area-matched, *i.e.*

$$\int_0^{\frac{TE}{2}} \vec{g}^{(app)}(t)dt = \int_{\frac{TE}{2}}^{TE} \vec{g}^{(app)}(t)dt \qquad \text{(Eq. C4.5-5)}$$

At the spin-echo time, the phase factor of the transverse magnetization (Eq. C4.5-1) becomes position independent and dephasing is zero.

Furthermore, for the case of a constant gradient only (*i.e.* $\vec{g}^{(app)}(t) = 0$ and $\vec{g}(t) = \vec{g}_0$) we obtain the familiar spin-echo result:

$$m(\vec{x}, t) = m(\vec{x}, 0) \exp[-i\omega_0 t] \exp\left[-\frac{t}{T_2}\right] \exp\left[-\frac{(2\pi\gamma)^2 |\vec{g}_0|^2 D t^3}{12}\right]$$

$$\times \exp[-i2\pi\vec{x} \cdot (\vec{k}(t) - 2H(t-\tau)\vec{k}(\tau))] \qquad \text{(Eq. C4.5-6)}$$

which when compared to the free decay equation Eq. 4.4-12, exemplifies several benefits of the spin-echo pulse sequence, specifically: 1) dephasing reversal or rephasing, and 2) reduced diffusive echo attenuation because of the factor 12 in the denominator of the diffusion attenuation factor.

## C4.6. *Spin-echo: Effects of Anisotropic Diffusion and Flow*

In this section we study the spin-echo experiment for spin packets with anisotropic diffusion and flow. The temporal evolution of the transverse magnetization is dictated by the BTS equation:

$$\frac{\partial m}{\partial t} = -i2\pi\gamma B_z^{(Exp)}m - \frac{m}{T_2} + i2\pi\gamma B_{1T}^{(Exp)}M_z + \vec{\nabla} \cdot \mathbf{D} \cdot \vec{\nabla}_m - \vec{v} \cdot \vec{\nabla}_m$$

$$\text{(Eq. C4.6-1)}$$

and again we are interested in an ideal applied magnetic field of the form:

$$B_z^{(App)}(\vec{x}, t) = B_0 + \vec{x} \cdot \vec{g}(t) \qquad \text{(Eq. C4.6-2)}$$

This mathematical problem was originally studied by Stejskal in 1965 (Stejskal, 1965) and was further refined by Basser *et al.* (Basser, Mattiello, & LeBihan, 1994) who introduced the concept of effective diffusion tensor. The BTS equation solution for the complex transverse magnetization is,

$$m(\vec{x}, t) = m(\vec{x}, 0) \exp[-i\omega_0 t] \exp\left[-\frac{t}{T_2}\right] \exp\left[-\sum_{l,m=x,y,z} D_{l,m}^{(eff)} b_{l,m}\right]$$

$$\times \exp[i\Phi(\vec{x}, t)] \exp[iH(t-\tau)(2\varphi - \pi)] \qquad \text{(Eq. C4.6-3)}$$

where the effective diffusion tensor (of rank 2) is the mean of the diffusion tensor over the time interval [0-TE], *i.e.*

$$\mathbf{D}^{(\text{eff})}(\vec{x}) \equiv \frac{1}{\text{TE}} \int_0^{\text{TE}} \mathbf{D}(\vec{x}, t) dt \qquad \text{(Eq. C4.6–4)}$$

Accordingly, the components of the effective diffusivity tensor are time-averaged quantities that are not explicit functions of time. The b-factor of the pulse sequence is defined as

$$b(t) = 2(\pi)^2 \left\{ \int_0^t [\vec{k}(t') - 2H(t - \tau)\vec{k}(\tau)\vec{k}(\tau)]^T \right.$$

$$\left. \times [\vec{k}(t') - 2H(t - \tau)\vec{k}(\tau)] dt' \right\} \qquad \text{(Eq. C4.6–5)}$$

The exponent of the echo attenuation factor is given by

$$\sum_{l,m=x,y,z} D_{l,m}^{(\text{eff})} b_{l,m} = (2\pi)^2 \left\{ \int_0^t [\vec{k}(t') - 2H(t - \tau)\vec{k}(\tau)]^T \cdot \mathbf{D}^{(\text{eff})} \right.$$

$$\left. \cdot [\vec{k}(t') - 2H(t - \tau)\vec{k}(\tau)] dt' \right\} \qquad \text{(Eq. C4.6–6)}$$

or equivalently,

$$\sum_{l,m=x,y,z} D_{l,m}^{(\text{eff})} b_{l,m} = D_{xx}^{(\text{eff})} b_{xx} + 2D_{xy}^{(\text{eff})} b_{xy} + 2D_{xz}^{(\text{eff})} b_{xz}$$

$$+ D_{yy}^{(\text{eff})} b_{yy} + 2D_{yz}^{(\text{eff})} b_{yz} + D_{zz}^{(\text{eff})} b_{zz}$$

$$\text{(Eq. C4.6–7)}$$

In addition, Stejskal's solution for the phase of the transverse magnetization is

$$\Phi(\vec{x}, t) = -2\pi \left\{ (\vec{x} - \vec{S}) \cdot (\vec{k}(t) - 2H(t - \tau)\vec{k}(\tau)) \right.$$

$$\left. + \gamma \int_0^t \vec{S} \cdot \vec{g} dt' - 2\gamma H(t - \tau) \int_0^\tau \vec{S} \cdot \vec{g} dt' \right\} \qquad \text{(Eq. C4.6–8)}$$

where we have defined the displacement vector

$$\vec{S}(t) \equiv \int_0^t \vec{v}(t')dt' \qquad \text{(Eq. C4.6–9)}$$

which is the shift in position of a spin packet during the time interval $[0,t]$ and, as before, the k-vector of the applied gradient is:

$$\vec{k}(t) \equiv \gamma \int_0^t \vec{g}(t')dt' \qquad \text{(Eq. C4.6–10)}$$

In summary, the presented BTS equation solution has several remarkable properties: 1) diffusion affects only the magnitude of the magnetization, *i.e.* echo amplitude reduction, 2) flow results in a phase shift only, and 3) the flow phase shift is not necessarily refocused at the time of the spin-echo.

## C5. The MR Imaging Scale: Voxels

### C5.1. *The MRI Pixel Value Equation: 3D-FT Imaging*

Signal measurement in MRI generates a series of complex numbers that are proportional to the volume integral of the complex transverse magnetization distribution multiplied by, or effectively weighted by the detection sensitivity function $(\Omega(\vec{x}))$ of the receiving coil, at each point of the imaging volume. Furthermore, every echo-signal-measurement $(S_{(i,j,sl)})$ is performed at a certain time after the rf excitation event $(te_{(i,j,sl)})$, each of which defines a different point of a grid in signal space, accordingly:

$$S_{(i,j,sl)} = \underset{\left\{\begin{smallmatrix}\text{Infinite}\\\text{space}\end{smallmatrix}\right\}}{\int\int\int} \Omega(\vec{x})m^{(A)}(\vec{x}, te_{(i,j,sl)})d^3x \qquad \text{(Eq. C5.1–1)}$$

The theoretical connection between spatial encoding with magnetic field gradients and the Fourier transform formalism can be established by solving the Bloch-Torrey-Stejskal equations with gradient pulses, as shown in sections C4.4 and C4.5. MR images are generated by inverse Fourier transforming the k-space arrays of encoded signals and the resulting pixel values are calculated with the equation:

$$pv_{(i,j,sl)} = \frac{1}{N_p \; N_f \; N_{sl}} \sum_{p,f,sl=0}^{N} S_{(f,p,sl)} \exp[+2\pi i \vec{X}_{(i,j,sl)} \cdot \vec{K}_{(f,p,sl)}]$$

$$\text{(Eq. C5.1–2)}$$

The array of pixel values is labeled by the three vertex coordinates of the imaging grid in geometric space, which is defined *via* the array of grid position vectors:

$$\vec{X}_{(i,j,sl)} = i\Delta x_{fe}\hat{f} + j\Delta x_{pe}\hat{p} + sl\Delta x_{ss}\hat{s} \qquad \text{(Eq. C5.1–3)}$$

where $\Delta x_{fe}$, $\Delta x_{pe}$, and $\Delta x_{ss}$ are the grid spacing along the frequency encoding (fe), phase encoding (pe), and slice selection direction respectively. These grid vectors are special cases of the continuous vector position, specifically:

$$\vec{x} = x_{fe}\hat{f} + x_{pe}\hat{p} + x_{ss}\hat{s} \qquad \text{(Eq. C5.1–4)}$$

where $(\hat{f}, \hat{p}, \hat{s})$ is the set of unitary vectors defining the three orthogonal directions of the geometric space as defined by the imaging gradients, and do not necessarily coincide with the laboratory axes $(\hat{x}, \hat{y}, \hat{z})$. Note that from now on, we will use the symbol "sl" instead of the usual "k" for designating the third index (slice index) of the imaging grid; this choice is convenient to avoid confusion with the coordinates in Fourier domain (k-space). Furthermore, the three integers (i, j, sl) are used as labels for the grid positions along the frequency encoding, phase encoding, and slice selection directions. The values of these imaging indices are in the ranges $\{0, N_\alpha - 1\}$ where $\alpha = $ fe, pe, sl respectively and $N_\alpha$ are the matrix sizes of the acquired three-dimensional array. The corresponding imaging grid in Fourier domain (signal space) is given by the raster vector:

$$\vec{K}_{(i,j,sl)} = \left(i - \frac{(N_{fe} - 1)}{2}\right)\Delta k_{fe}\hat{f} + \left(j - \frac{(N_{pe} - 1)}{2}\right)\Delta k_{pe}\hat{p}$$

$$+ \left(sl - \frac{(N_{sl} - 1)}{2}\right)\Delta k_{ss}\hat{s} \qquad \text{(Eq. C5.1–5)}$$

where the conjugate spacing relationships of discrete Fourier mathematics apply:

$$\Delta x_\alpha \Delta k_\alpha = \frac{1}{N_\alpha}; \quad \alpha = \text{fe, pe, ss} \qquad \text{(Eq. C5.1–6)}$$

and the k-space is centered about $\vec{K} = \vec{0}$ in three dimensions.

By generalizing the BTS solutions of sections C4.4-6 to arbitrary pulse sequences, we can write the following equation

$$m^{(A)}(\vec{x}, te_{(f,p,sl)}) = \left(\frac{\pi^2 \gamma^2 \hbar^2 B_0}{k_B T}\right) PD^{(A)}(\vec{x}) psw(qCV||qNMR\_par^{(A\&B)}(\vec{x}))$$

$$\times \exp\left(-\frac{te_{(f,p,sl)}}{T_2^{(A)}(\vec{x})}\right) \exp(-2\pi i \vec{x} \cdot \vec{K}_{(f,p,sl)}) \quad \text{(Eq. C5.1-7)}$$

which expresses the transverse magnetization phasor in terms of several factors including the proton density, a generic reduced pulse sequence weighting factor, the exponential $T_2$ decay factor, and the Fourier factor. The reduced pulse sequence weighting factor models the evolution of the transverse magnetization during a pulse sequence and includes all dynamical effects other than $T_2$ decay, which because of its importance, has been expressed separately and explicitly. The reduced pulse sequence factor is a function of many timing parameters and gradient amplitudes: in Eq. C5.1-7, we have made explicit only the quantitative control variables (qCV), which play a fundamental role in qMRI because their values control the degree of weighting to a given qMRI parameter. qCV examples include TR and TE for controlling the degrees of $T_1$- and $T_2$-weighting respectively and many other exist, as will be discussed in Chapter E.

The specific mathematical form of the pulse sequence weighting factor must be derived by solving the appropriate BTS equations and is in general complex-valued, accordingly we write

$$psw(qCV||qNMR\_par^{(A\&B)}(\vec{x})) = |psw| \exp(i\Phi_{psw}) \quad \text{(Eq. C5.1-8)}$$

Both, the magnitude and the phase of the pulse sequence weighting factor can be functions of pulse sequence control variables and the qNMR parameters of the tissues and its $^1$H-proton pools (liquid and semisolid pools). The phase factor includes all phases accumulated by the spin packet's transverse magnetization during a pulse sequence cycle, with the exception of the phase intended for spatial encoding. Some contributions to this phase will result from the inevitable magnetic field inhomogeneities and other imperfections in the imaging experiment, and some can be intentionally generated with applied gradients; as will be shown later, this is done typically for quantifying flow or motion. The pulse sequence weighting factor depends functionally on the pulse sequence type, the extent of magnetization saturation depending on TR, as well as the effects of any applied pulses affecting the decay of the transverse magnetization through each pulse sequence cycle. We shall see later in Chapter E that qMRI accuracy will depend on several factors and in particular, on how faithfully the pulse sequence weighting factors models the pulse sequence.

By inserting Eq. C5.1-7 into Eq. C5.1-1, then substituting the subsequent equation into Eq. C5.1-2, and finally after reordering factors, we derive the following pixel value equation,

$$
pv_{(i,j,sl)} = \left( \frac{\pi^2 \gamma^2 \hbar^2 B_0}{k_B T} \right) \underset{\left\{ \begin{array}{c} \text{Infinite} \\ \text{space} \end{array} \right\}}{\int \int \int} \left\{ \Omega(\vec{x}) PD^{(A)}(\vec{x}) psw \right.
$$

$$
\times \, (qCV \| qNMR\_par^{(A\&B)}(\vec{x}) \exp \left( -\frac{TE}{T_2^{(A)}(\vec{x})} \right) \Bigg\}
$$

$$
\times VSF^{(3D)}(\vec{X}_{(i,j,sl)} - \vec{x}) d^3 x \qquad \text{(Eq. C5.1–9)}
$$

in which, the echo time TE has been defined as the time at the center of the readout window, and the voxel sensitivity function (VSF) of 3D imaging is given by

$$
VSF^{(3D)}(\vec{X}_{(i,j,sl)} - \vec{x}) \equiv \frac{1}{N_p \, N_f \, N_{sl}} \sum_{p,f,sl=0}^{N} \exp \left( \frac{TE - te_{(f,p,sl)}}{T_2^{(A)}(\vec{x})} \right)
$$

$$
\times \, \exp[+2\pi i (\vec{X}_{(i,j,sl)} - \vec{x}) \cdot \vec{K}_{(f,p,sl)}] \quad \text{(Eq. C5.1–10)}
$$

As illustrated in Fig. B-2 in two dimensions, this function is highly localized around each grid point in geometric space. This equation therefore provides the theoretical foundation of spatial localization, and therefore imaging. It is also a function of $T_2$ and becomes broader in the short $T_2$ limit, *i.e.* for $T_2(\vec{x}) < TE - te_{(f,p,sl)}$. On the other hand, in the long $T_2(\vec{x}) \gg TE - te_{(f,p,sl)}$ regime, the voxel sensitivity function becomes $T_2$-independent and furthermore, it can be expressed in analytical form (Parker, Du, & Davis, 1995):

$$
VSF^{(3D)}(\vec{X}_{(i,j,sl)} - \vec{x}) = \frac{1}{N_f N_p N_{sl}} \frac{\sin \left( \frac{\pi (X_{(i,j,sl)} - x)}{\Delta x} \right)}{\sin \left( \frac{\pi (X_{(i,j,sl)} - x)}{N_f \Delta x} \right)} \frac{\sin \left( \frac{\pi (Y_{(i,j,sl)} - y)}{\Delta y} \right)}{\sin \left( \frac{\pi (Y_{(i,j,sl)} - y)}{N_p \Delta y} \right)}
$$

$$
\times \frac{\sin \left( \frac{\pi (Z_{(i,j,sl)} - z)}{\Delta z} \right)}{\sin \left( \frac{\pi (Z_{(i,j,sl)} - z)}{N_f \Delta z} \right)} \qquad \text{(Eq. C5.1–11)}
$$

where we have arbitrarily assigned the x-direction as the frequency encoding direction and the two phase encoding directions are in the y- and z-directions.

The voxel volume is equal to the volume integral of the voxel sensitivity function:

$$\Delta V = \underset{\left\{\substack{\text{Infinite}\\\text{space}}\right\}}{\int\int\int} \mathrm{VSF}^{(3D)}(\vec{X}_{(i,j,sl)} - \vec{x})d^3x = \Delta x \Delta y \Delta z$$

(Eq. C5.1–12)

and equals the product of the pixel spacings in the long $T_2$ regime.

## C5.2. *The MRI Pixel Value Equation: 2D-FT Imaging*

For 2D pulse sequences, a nearly identical pixel value equation holds, specifically:

$$pv_{(i,j,sl)} = \left(\frac{\pi^2 \gamma^2 \hbar^2 B_0}{k_B T}\right) \underset{\left\{\substack{\text{Infinite}\\\text{space}}\right\}}{\int\int\int} \left\{ \Omega(\vec{x}) PD^{(A)}(\vec{x}) psw \right.$$

$$\times (qCV \| qNMR\_par^{(A\&B)}(\vec{x})) \exp\left(-\frac{TE}{T_2^{(A)}(\vec{x})}\right)\Bigg\}$$

$$\times VSF^{(2D)}(\vec{X}_{(i,j,sl)} - \vec{x})d^3x \qquad \text{(Eq. C5.2–1)}$$

except that the 2D voxel sensitivity function has Fourier localization only along the two in-plane directions and consists of the slice profile in the third dimension as results from the slice selective rf pulses. Accordingly,

$$VSF^{(2D)}(\vec{X}_{(i,j,sl)} - \vec{x}) \equiv \frac{S(X_{sl} - x_{ss})}{N_p\,N_f} \sum_{p,f=0}^{N} \exp\left(\frac{TE - te_{(f,p)}}{T_2^{(A)}(\vec{x})}\right)$$

$$\times \exp[+2\pi i(\vec{X}_{(i,j)} - \vec{x})\cdot\vec{K}_{(f,p)}] \quad \text{(Eq. C5.2–2)}$$

where $S(X_{sl} - x_{sl})$ is the slice profile that results from using slice-selective rf pulses, as discussed in section C4.1. As in the 3D case, the voxel sensitivity function is $T_2$ dependent in the short $T_2$ regime and becomes $T_2$-independent in the long $T_2$ regime, in which it can be expressed in analytical form, specifically:

$$\text{VSF}^{(2D)}(\vec{X}_{(i,j,sl)} - \vec{x}) = \frac{S(X_{sl} - x_{ss})}{N_f \, N_p} \frac{\sin\left(\frac{\pi(X_{(i,j,sl)} - x)}{\Delta x}\right)}{\sin\left(\frac{\pi(X_{(i,j,sl)} - x)}{N_f \Delta x}\right)}$$

$$\times \frac{\sin\left(\frac{\pi(Y_{(i,j,sl)} - y)}{\Delta y}\right)}{\sin\left(\frac{\pi(Y_{(i,j,sl)} - y)}{N_p \Delta y}\right)} \qquad \text{(Eq. C5.2–3)}$$

Furthermore, in this case, the 2D grids in anatomic and k-space are independent of the slice index, specifically

$$\vec{X}_{(i,j)} = i\Delta x_{fe}\hat{f} + j\Delta x_{pe}\hat{p} \qquad \text{(Eq. C5.2–4)}$$

and the k-space trajectory is on a slice-by-slice basis and given by:

$$\vec{K}_{(i,j)} = \left(i - \frac{(N_{fe} - 1)}{2}\right)\Delta k_{fe}\hat{f} + \left(j - \frac{(N_{pe} - 1)}{2}\right)\Delta k_{pe}\hat{p}$$

$$\text{(Eq. C5.2–5)}$$

### C5.3. *Pixel Value Equation: Exact Integral Form*

The main results of the previous two sections specifically Eq. C5.1–9 and Eq. C5.2–1 model MRI pixel values as volume integrals in which the integrands --inside curly parentheses-- are weighted by, or "localized" by, the appropriate 2D or 3D voxel sensitivity function depending on the spatial encoding scheme of the pulse sequence used. We rewrite these two equations using the common notation,

$$pv_{(i,j,sl)} = \left(\frac{\pi^2 \gamma^2 \hbar^2 B_0}{k_B T}\right) \iiint_{\left\{\substack{\text{Infinite}\\ \text{space}}\right\}} \left\{\Omega(\vec{x})PD^{(A)}(\vec{x})|psw|\exp(i\Phi_{psw})\right.$$

$$\left. \times \exp\left(-\frac{TE}{T_2^{(A)}(\vec{x})}\right)\right\} \text{VSF}(\vec{X}_{(i,j,sl)} - \vec{x})d^3x \qquad \text{(Eq. C5.3–1)}$$

with the understanding that the appropriate 2D- or 3D-VSF should be used when analyzing images acquired with 2D and 3D pulse sequences respectively. In addition, the pulse sequence weighting factor has been expressed in terms of its modulus and phase factor.

The pixel value equation above is still not in a form that can be readily used for qMRI processing, which requires performing the volume integral thus providing a mathematical venue for migrating from the qNMR parameters of spin packet to the qMRI parameters of imaging voxels.

Transitioning from the spatial scale of spin packets to the imaging-voxel scale is a delicate theoretical undertaking since the architecture of tissues is strongly dependent on the spatial scale considered. Because of their small size, spin packets are relatively simple moieties that are therefore well approximated by the two-pool NMR model discussed above. On the other hand, at the voxel scale tissues exhibit intricate and diverse biological morphology with distinct geometrical features including intracellular organelles, cell membranes, cells, cell clusters, intra- and extra-cellular spaces, microfibers, the microvascular bed, and so forth. Furthermore, when transitioning to the imaging-voxel scale, qMRI parameters can be defined that do not have qNMR correlates, specifically: 1) diffusional anisotropy parameters for tissues with organized micro-architecture such that diffusion is preferential along a given direction(s). 2) Functional parameters associated with blood microcirculation --tissue perfusion qMRI parameters--, and 3) $T_2^*$, an additional qMRI parameter that incorporates the dephasing effect of voxel-scale magnetic field inhomogeneities.

At the voxel scale, biologic tissue has been modeled as consisting of a perfusion volume fraction $f^{(A)}$ (typically a few percent) of blood water flowing in the randomly oriented vessels of the capillary network, and a fraction $(1-f^{(A)})$ of the "immobile" (diffusing only) intra- and extracellular water. In this model, which is often referred to as (Turner, Le Bihan, & Chesnick, 1991; Turner *et al.*, 1990) Intra-Voxel Incoherent Motion (IVIM), a new qMRI parameter arises, specifically a perfusion related diffusion coefficient $D^{(A)^*}$ of the capillary network, which is generally larger that the diffusion of the static free pool $D^{(A)}$.

We can infer an approximate form of the pixel value equation, which is particularly suited for developing qMRI algorithms by assuming that the qNMR tissue parameters do not vary significantly within a voxel and that therefore qMRI parameters may be interpreted as voxel-averaged qNMR parameters, *i.e.*:

$$\text{qMRI\_par}_{(i,j,sl)} = \langle \text{qNMR\_par}(\vec{x}) \rangle_{(i,j,sl)} \qquad \text{(Eq. C5.3-2)}$$

where the bracket notation denotes average over the voxel. We emphasize however that, as discussed above, not all qMRI parameters have qNMR analogues, specifically the parameters of diffusional anisotropy, perfusion, and intravoxel dephasing; these become well-defined parameters only at the larger spatial scale of voxels.

With these considerations in mind, we can write an approximate form of the pixel value equation:

$$pv_{(i,j,sl)} = \left(\frac{\pi^2\gamma^2\hbar^2 B_0}{k_B T}\right)\Omega_{(i,j,sl)}PD_{(i,j,sl)}^{(A)}|psw(qMRI\_par_{(i,j,sl)})|$$

$$\times \exp\left(-\frac{TE}{T_{2(i,j,sl)}^{(A)}}\right)\exp(-i2\pi\gamma B_{0,(i,j,sl)}TE)I_{(i,j,sl)}(TE)$$

(Eq. C5.3–3)

For this equation, we have defined the intravoxel dephasing integral:

$$I_{(i,j,sl)}(TE) \equiv \iiint\limits_{\left\{\substack{\text{Infinite}\\\text{space}}\right\}} \exp(i\Phi_w^{(Inh)}(\vec{x}, TE))VSF(\vec{X}_{(i,j,sl)} - \vec{x})d^3x$$

(Eq. C5.3–4)

Furthermore, we have transitioned from qNMR parameters to qMRI parameters as arguments of the pulse sequence weighting factor modulus. We left the phase factor inside the integral sign because the inhomogeneity magnetic field will in general exhibit spatial variations at the subvoxel scale.

The inhomogeneity magnetic field is the main source of dephasing in MRI. As discussed earlier, it arises from the inevitable magnetic susceptibility differences within the imaged object. Other possible causes for dephasing are hardware nonlinearities and imperfections.

## C5.4. *Voxel-Integrated Pixel Value Equation: Spin-echo*

For spin-echo pulse sequences, the phase accrued due to the inhomogeneity field vanishes at the echo time $\Phi_w^{(Inh)}(\vec{x}, TE) = 0$ and the pixel value equation is simply:

$$pv_{(i,j,sl)} = \left(\frac{\pi^2\gamma^2\hbar^2 B_0}{k_B T}\right)\Omega_{(i,j,sl)}PD_{(i,j,sl)}^{(A)}|psw(qMRI\_par_{(i,j,sl)})|$$

$$\times \exp\left(-\frac{TE}{T_{2,(i,j,sl)}^{(A)}}\right)\Delta V_{(i,j,sl)} \qquad \text{(Eq. C5.4–1)}$$

where we have made use of the equality:

$$\Delta V_{(i,j,sl)} = \underset{\left\{\substack{\text{Infinite}\\\text{space}}\right\}}{\int\int\int} \text{VSF}(\vec{X}_{(i,j,sl)} - \vec{x})d^3x \qquad \text{(Eq. C5.4-2)}$$

We note that the voxel volume is equal to the product of the pixel spacing along each direction, only in the long $T_2$-regime. The pixel value equation above is the basis for developing qMRI algorithms for spin-echo pulse sequences.

## C5.5. *Voxel-Integrated Pixel Value Equation: Gradient-Echo*

For gradient-echo pulse sequences, the situation is considerably more complicated because the inhomogeneity phase dispersion is not refocused and therefore we have to deal with the voxel sensitivity function integral of the intravoxel varying phase, specifically:

$$I_{(i,j,sl)}(\text{TE}) \equiv \underset{\left\{\substack{\text{Infinite}\\\text{space}}\right\}}{\int\int\int} \exp(i\Phi_w^{(\text{Inh})}(\vec{x}, \text{TE}))\text{VSF}(\vec{X}_{(i,j,sl)} - \vec{x})d^3x$$

$$\text{(Eq. C5.5-1)}$$

In other words, for gradient-echo pulse sequences the process of voxel formation remains entangled with magnetization dynamics in a manner that not only leads to accelerated $T_2$ relaxation but also such that it can interfere with proper voxel formation. For this reason, gradient-echo pulse sequences are also prone to geometric deformation artifacts and signal dropouts in regions of magnetic field inhomogeneity, particularly for long TEs.

To complicate matters even further, the inhomogeneity field is generally unknown. In the following, we will analyze the pixel value integral above assuming an inhomogeneity magnetic field of the form:

$$\delta\vec{B}_0(\vec{x}) = g^{(\text{Inh})}z\hat{z} \qquad \text{(Eq. C5.5-2)}$$

and therefore,

$$\Phi_{psw}^{\text{Inh}}(\vec{x}, \text{TE}) = 2\pi\gamma g^{(\text{Inh})}z\text{TE} \qquad \text{(Eq. C5.5-3)}$$

In this case the integral is:

$$I_{(i,j,sl)}(TE) \equiv \underset{\left\{\begin{smallmatrix} \text{Infinite} \\ \text{space} \end{smallmatrix}\right\}}{\int\int\int} \exp(i2\pi\gamma g^{(Inh)}z\,TE)VSF(\vec{X}_{(i,j,sl)} - \vec{x})d^3x$$

(Eq. C5.5–4)

By recalling the 2D-VSF equation, specifically:

$$VSF^{(2D)}(\vec{X}_{(i,j,sl)} - \vec{x}) = \frac{S(X_{sl} - x_{ss})}{N_f N_p} \frac{\sin\left(\pi\frac{(X_{(i,j,sl)} - x)}{\Delta x}\right)}{\sin\left(\pi\frac{(X_{(i,j,sl)} - x)}{N_f \Delta x}\right)}$$

$$\times \frac{\sin\left(\pi\frac{(Y_{(i,j,sl)} - y)}{\Delta y}\right)}{\sin\left(\pi\frac{(Y_{(i,j,sl)} - y)}{N_p \Delta y}\right)} \qquad \text{(Eq. C5.5–5)}$$

and considering that for 2D acquisitions, the largest dimension of the voxel is along the slice encoding direction, a model can be derived, as an exponential decay factor multiplied by TE-dependent voxel volume:

$$I_{(i,j,sl)}(TE) \approx \exp\left(-\frac{TE}{T_{2,(i,j,sl)}^{(A\_Inh)}}\right) \text{sinc}\left(2\pi\gamma\frac{\Delta B_0}{2}TE\right)\Delta V_{(i,j,sl)}$$

(Eq. C5.5–6)

where $\Delta B_0$ is the variation of the magnetic field along the slice-select voxel dimension and

$$\frac{1}{T_2^{(A\_Inh)}} = \gamma\Delta B_0 \qquad \text{(Eq. C5.5–7)}$$

This approximates the reduced transverse relaxation time as the inverse of the absorption linewidth. Here we have used a different symbol for this intravoxel-dephasing relaxation time, which is commonly denoted by $T_2$ prime.

Hence, assuming that a voxel free induction decay (FID) signal decays exponentially in time is a crude approximation in most cases; nevertheless, this assumption leads to a widely used qMRI parameter, which is known as the reduced transverse relaxation time ($T_2^*$). This voxel size dependent parameter combines in one parameter the two main causes of voxel-FID decay, specifically the (average) intermolecular $T_2$ decay or natural $T_2$ decay, and the magnetic field inhomogeneity $T_2$ decay. Accordingly,

$$\frac{1}{T_2^{(A)*}} = \frac{1}{T_2^{(A)}} + \frac{1}{T_2^{(A\_Inh)}} \qquad \text{(Eq. C5.5–8)}$$

In liquids and tissues, the major factor influencing voxel-FID $T_2$ decay is the magnetic field inhomogeneity within the voxel ($\Delta B_0$). For example, for brain gray matter at 1.5T, a 0.5ppm inhomogeneity across the voxel will lead to $T_2^{(A\_Inh)} \cong 31\,\text{ms}$; since $T_2^{(A)} \cong 100\,\text{ms}$ we find $T_2^{(A)*} \cong 24\,\text{ms}$

We note in the limit of perfect magnetic homogeneity ($\Delta B_0 \longrightarrow 0$), the inhomogeneity phase vanishes ($\Phi_w^{(Inh)}(\vec{x}, TE) \longrightarrow 0$), $T_2^{(A\_Inh)} \longrightarrow \infty$, and the voxel-FID is dominated by the intermolecular $T_2$ decay. If on the other hand $\Delta B_0 \neq 0$, then the inhomogeneity phase can still be made zero if a spin-echo is acquired (see section C4.5), and the pixel value equation above still applies.

In conclusion, pixel values for gradient-echo pulse sequences are modeled by:

$$pv_{(i,j,sl)} = \left(\frac{\pi^2 \gamma^2 \hbar^2 B_0}{k_B T}\right) \Omega_{(i,j,sl)} PD_{(i,j,sl)}^{(A)} |psw(qMRI\_par_{(i,j,sl)})|$$

$$\times \exp\left(-\frac{TE}{T_{2(i,j,sl)}^{(A)*}}\right) \text{sinc}\left(2\pi\gamma \frac{\Delta B_0}{2} TE\right)$$

$$\times \exp(-i2\pi\gamma B_{0,(i,j,sl)} TE)\Delta V_{(i,j,sl)} \qquad \text{(Eq. C5.5–9)}$$

which is very similar to the spin-echo pixel value equation with $T_2^*$ replacing $T_2$. It also has an extra phase factor that represents the relative dephasing from $B_0$ inhomogeneities at larger inter-voxel scale. Other differences between gradient-echo and spin-echo sequences relate to differences in the modulus of the pulse sequence weighting factor in relation to $T_1$ dependencies (magnetization saturation properties) and diffusion coefficient dependencies.

## C5.6. *Classification of qMRI Parameters*

The main objective of qMRI is to map the spatial distributions of all the qMRI parameters in Eq. C5.5-9, and this form of the pixel value equation is instrumental for developing qMRI techniques. On one hand, it provides theoretical guidelines for designing qMRI pulse sequences and on the other hand, because it has a multiplicative structure, it is ideally suited for deriving the mathematical formulas at the core of qMRI algorithms, as will be shown in Chapter E.

As in the case of spin packets (see section C3.9) and adopting now a "voxel-scale two-pool model", the qMRI parameters can be organized in similar tabular arrangements as the qNMR parameters, *i.e.*

$$
\text{qMRI\_par}^{(A)} = \left\{ \begin{array}{cccc} \text{PD}^{(A)} & \ldots & \ldots & \ldots \\ \vdots & \mathbf{D}^{(A)} & (f^{(A)}, D^{*(A)}) & \vec{v}^{(A)} \\ \vdots & \text{T}_1^{(A)} & (\text{T}_2^{(A)}, \text{T}_2^{*(A)}) & \text{T}_{1,2\rho}^{(A)} \end{array} \right\}
$$

(Eq. C5.6–1)

where, the first, second, and third rows consist of the spin densities, the kinetic parameters, and the MR interaction parameters, respectively. We emphasize again that the additional parameters associated to perfusion $(f^{(A)}, D^{(A)}*)$ and dephasing $\text{T}_2^{(A)*}$, do not have qNMR correlates.

For the semisolid pool, which is coupled to the liquid pool *via* the fundamental decay rate (R), we write:

$$
\text{qMRI\_par}^{(B)} = \left\{ \begin{array}{cccc} \text{PD}^{(B)} & \ldots & \ldots & \ldots \\ \vdots & \mathbf{D}^{(B)} = 0 & (f^{(B)}, D^{(B)*}) = 0 & \vec{v}^{(B)} = \vec{v}^{(A)} \\ \vdots & \text{T}_1^{(B)} & \text{T}_2^{(B)*}, \text{T}_2^{(B)} & \text{T}_{1,2\rho}^{(B)} \end{array} \right\}
$$

(Eq. C5.6–2)

Finally, for the sake of completeness of this section, we rewrite the relationships between spin densities and the measured equilibrium longitudinal magnetizations of both pools, specifically:

$$
\text{M}_Z^{(A\_eq)} \cong \left( \frac{\pi^2 \gamma^2 \hbar^2 B_0}{k_B T} \right) \text{PD}^{(A)}
$$

(Eq. C5.6–3)

and,

$$
\text{M}_Z^{(B\_eq)} \cong \left( \frac{\pi^2 \gamma^2 \hbar^2 B_0}{k_B T} \right) \text{PD}^{(B)}
$$

(Eq. C5.6–4)

Altogether, the qMRI parameters of Eq. C5.6-1 through Eq. C5.6-4 form the family of primary qMRI parameters for the voxel-scale two-pool model of biologic tissue. The accuracy of such an idealized tissue model increases as a function of increasing spatial resolution and in the ultra high spatial resolution limit in which the voxel volume approaches the spin packet volume, i.e. $\Delta V \longrightarrow \delta V$, we have in general:

$$\text{qMRI\_par} \xrightarrow{\Delta V \longrightarrow \delta V} \text{qNMR\_par} \qquad \text{(Eq. C5.6–5)}$$

and also,

$$T_2^{(A)*} \xrightarrow{\Delta V \longrightarrow \delta V} T_2^{(A)} \qquad \text{(Eq. C5.6–6)}$$

Hence, the ultrahigh spatial resolution limit could be referred also as the NMR imaging limit.

## C6. MRI Pulse Sequences

### C6.1. *Basic Concepts*

Spatial localization in MRI is achieved by means of magnetic field gradients, which are therefore also known as imaging gradients. As discussed throughout this book, these are time- and spatially-dependent applied magnetic fields that are parallel to the polarizing magnetic field --*i.e.* the laboratory z-direction-- and that have the defining property that their magnitudes vary linearly along any given spatial direction. By using three such magnetic field gradients with spatial variations along three orthogonal directions, NMR signal localization in three dimensions can be achieved. MRI pulse sequences are streams of rf pulses and magnetic field gradient pulses. The pulses are transmitted in four separate channels -one rf signal generation channel and three spatial encoding channels- and are timed according to rules compatible with: NMR physics, spatial encoding physics, and the desired scan properties (contrast weighting, spatial resolution, anatomic coverage, image orientation, artifact reduction techniques).

An MRI pulse sequence is a cyclical and semi-repetitive series of NMR experiments with which the sample's magnetization is driven out of equilibrium by a series of brief bursts of resonant rf energy and pulsed magnetic field perturbations (position-tagging gradient pulses). The duration of each pulse sequence cycle is known as the repetition time and is universally denoted by TR. Every cycle of the pulse sequence begins with an rf pulse, the purpose of which is to initiate the NMR signal and is therefore referred to as the excitation pulse. The NMR signal measurement is performed a certain time after the excitation pulse, at an operator-adjustable time that is referred to as the echo time TE. The position-tagging pulses are applied between the time of excitation and TE.

The scan time for a single slice with $N_{ph}$ phase encoding steps is:

$$SCT(1 \text{ slice}) = N_{ph} \text{ TR} \qquad \text{(Eq. C6.1–1)}$$

where $N_{ph}$ is the number of phase encoding steps.

### C6.2. *Pulse Sequence Classification*

Since rf and gradient pulses can be applied in innumerable combinations there is in principle infinite an number of possible MRI pulse sequences. The three most important criteria useful for categorizing FT-MRI pulse sequences relate to: 1) the spatial encoding scheme (scanning technique) used, 2) the type of NMR signal acquired, and 3) the data acquisition acceleration technique, whether one or several differently encoded signals are acquired per pulse sequence TR cycle. These criteria used for classifying pulse sequences are reviewed in the following sections.

### C6.3. *Spatial Encoding Schemes (Scanning Techniques)*

The purpose of an MRI acquisition is to generate a three-dimensional array of imaging data that spans a certain volume in three-dimensional space. Such an imaging volume may encompass the whole object or typically covers a certain subvolume of the object. The end product of applying a pulse sequence is an array of pixel values such that:

$$\{pv_{(i,j,sl)}\} \longleftarrow \begin{cases} i = 0, \ldots, N_{fe} - 1 \\ j = 0, \ldots, N_{pe} - 1 \\ sl = 0, \ldots, N_{ss} - 1 \end{cases} \qquad \text{(Eq. C6.3–1)}$$

FT-MRI affords several spatial encoding schemes, including: 1) multiple sequential 2D scanning (M2D), 2) multislice 2D scanning (MS2D), 3) single volume three-dimensional scanning (3D), and 4) multi-slab three-dimensional scanning (MS3D). With M2D, slices are interrogated one at a time and sequentially one after another. Consequently, the scan time for $N_{ss}$ slices is given by:

$$SCT^{(M2D)}(N_{ss} \text{ slices}) = N_{ss} \, N_{ph} \text{ TR} \qquad \text{(Eq. C6.3–2)}$$

As an example, the M2D scan time for 20 slices, 256-matrix size, at TR = 2s is 2.84 hours. Hence, M2D scan times are unacceptably long unless operating in the very short TR regime, which in turn translates into imaging with strong magnetization saturation.

As a much more time efficient alternative, with MS2D several different slices are interrogated within the TR interval. For a given TR, the total number of slices that can be interrogated without increasing the scan time is TR/TE and is known as a package of slices. For the example above, if TR = 2s and TE = 0.1s, a package will consist of 20 slices and the scan time, which is given by

$$SCT^{(MS\ 2D)}(1\ package) = N_{ph}\ TR \qquad (Eq.\ C6.3\text{--}3)$$

is only 8.5 minutes. MS2D thus allows for scanning a much-increased number of slices in acceptable scan times and without necessitating operation in the short-TR and regime.

With either of the 2D scanning modes above, Fourier encoding is used only along the two in-plane directions and not along the slice selection direction. With M2D and MS2D, the minimum slice thickness is inversely proportional to the maximum gradient strength of the imaging subsystem and typically limited to about 1 mm with current hardware (*e.g.* 30 mT/m). Much thinner slices can be obtained by using Fourier encoding along the slice selection direction. This is the case of 3D MRI pulse sequences, for which the scan time is given by,

$$SCT^{(3D)}(N_{ss}\ slices) = N_{ss}\ N_{ph}\ TR \qquad (Eq.\ C6.3\text{--}4)$$

which is mathematically identical to that of M2D scans, except that $N_{ss}$ is the matrix size of a Fourier encoding direction. Hence reasonable 3D scan times can be achieved in the very short-TR regime or by using a multi slab approach (MS3D) similar to the MS2D approach.

## C6.4. *Signal Types*

NMR signals are time dependent electrical currents induced in a receiver coil by the rotating transverse magnetizations of all the tissues in the imaging volume; MRI signals are spatially encoded NMR signals. The number of pulses in the rf channel of an MRI pulse sequence determines the NMR signal type that is generated. The simplest pulse sequences use only one rf pulse per TR cycle and generates an FID signal that decays rapidly in time. Next, the spin-echo pulse sequence uses two rf pulses, a 90° excitation followed by a 180° refocusing pulse, to generate spin-echo signals in which the dephasing effects of static magnetic field inhomogeneities have been

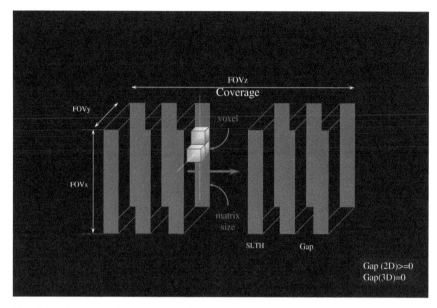

Fig. C-8.   Geometrical parameters of an MRI acquisition. The dimensions of the imaging volume are the field-of-views (FOVs) along each direction. The imaging dataset consists of Ns slices of slice thickness (SLTH). Each slice consist of Nx times Ny voxels and these are referred to as the matrix sizes.

compensated. The stimulated echo pulse sequence uses three 90° rf pulses. MRI pulse sequences using one rf excitation pulse per TR cycle are referred to as gradient-echo (GE) or field echo (FE) pulse sequences, denominations that are manufacturer specific. MRI pulse sequences based on the SE principle are referred to as conventional spin-echo (CSE) or simply spin-echo (SE). The acronym for the stimulated echo pulse sequence is STEAM.

## C6.5.  *Data Acquisition Acceleration Techniques*

### C6.5.i.  *Ultra-short TR Regime*

We have seen in the previous sections that scan time is proportional to TR and therefore, one obvious approach for achieving faster MRI scanning is reducing TR. Indeed some of the fastest and most time efficient MRI pulse sequences operate in the ultra-short TR regime in which TR is comparable to the $T_2$ of typical tissue in the imaging volume. Ultra-short TR pulse sequences (Oppelt *et al.*, 1986) (A. Haase, 1990) have typically one rf pulse per TR cycle — with a flip angle of less than 90° — thus resulting in train of identical rf pulses intertwined with gradient pulses. In this TR~$T_2$ regime, a steady state free precession (SSFP) signal can develop because the FID does not have sufficient time to decay in one TR cycle and if measures are taken for perfect refocusing of the spins, a steady state of both the longitudinal and the transverse can be achieved. Perfect refocusing is achieved by balancing the moments of all gradient pulses including the imaging gradient pulses (K. Scheffler & Hennig, 2001). As pointed out by Deoni *et al.* (Deoni, Rutt, & Peters, 2003), several SSFP pixel value equations have been presented in the literature that are valid for different variations of the SSFP pulse sequence and these authors adopt the variant published by Perkins and Wehrli (Perkins & Wehrli, 1986). This equation, which we reproduce here, is valid for very short TRs of about 10 ms or less and for rf pulses with alternating phases. Accordingly, and in the notation of this book:

$$\text{pv}_{(i,j,sl)} \cong \left( \frac{\pi^2 \gamma^2 \hbar^2 B_0}{k_B T} \right) \Omega_{(i,j,sl)} \text{PD}_{(i,j,sl)}^{(A)}$$

$$\times \frac{(1 - E_1^{(A)}(\text{TR})) \sin(\text{FA})}{1 - E_1^{(A)}(\text{TR}) E_2^{(A)}(\text{TR}) - (E_1^{(A)}(\text{TR}) - E_2^{(A)}(\text{TR})) \cos(\text{FA})}$$

$$\times \exp \left( -\frac{\text{TE}}{T_{2,(i,j,sl)}^{(A)*}} \right) \Delta V_{(i,j,sl)} \qquad \text{(Eq. C6.5–1)}$$

where we have further adopted the commonly used abbreviation for the exponential factors:

$$E_{1,2}^{(A)}(t) \equiv \exp \left( -\frac{t}{T_{1,2}^{(A)}} \right) \qquad \text{(Eq. C6.5–2)}$$

This pixel value equation is the basis for relaxometry with the SSFP pulse sequence, as will be reviewed in detail in the following sections of this chapter. Alternative equations can be found in a recent review article (K Scheffler & Lehnhardt, 2003).

### C6.5.ii. *Medium-Short TR Regime*

For longer TR values or if the transverse steady state magnetization is purposely spoiled, then we can take the limit $E_2^{(A)}(\text{TR}) \to 0$ and the gradient echo pixel value equation above takes a much simpler form, specifically:

$$pv_{(i,j,sl)} \cong \left(\frac{\pi^2\gamma^2\hbar^2 B_0}{k_B T}\right) \Omega_{(i,j,sl)} PD_{(i,j,sl)}^{(A)}$$

$$\times \frac{(1 - E_1^{(A)}(TR)) \sin(FA)}{1 - E_1^{(A)}(TR) \cos(FA)} \exp\left(-\frac{TE}{T_{2,(i,j,sl)}^{(A)*}}\right) \Delta V_{(i,j,sl)}$$

(Eq. C6.5–3)

In this case, only the longitudinal magnetization reaches a steady state and the $T_2$ coherences from TR cycle to TR cycle can be neglected and the pulse sequence is named fast low angle shot (FLASH) or fast field echo (FFE). This pixel value equation is the basis for $T_1$ relaxometry as described by Deoni *et al.* (Deoni, *et al.*, 2003).

Analogously, the spin-echo pixel value equation is:

$$pv_{(i,j,sl)} \cong \left(\frac{\pi^2\gamma^2\hbar^2 B_0}{k_B T}\right) \Omega_{(i,j,sl)} PD_{(i,j,sl)}^{(A)} \left(1 - \left(2\exp\left(-\frac{(TR - TE/2)}{T_1}\right)\right.\right.$$

$$\left.\left. -\exp\left(-\frac{TR}{T_1}\right)\right)\right) \exp\left(-\frac{TE}{T_{2,(i,j,sl)}^{(A)}}\right) \Delta V_{(i,j,sl)} \qquad \text{(Eq. C6.5–4)}$$

where we have assumed exact 90° and 180° pulses.

### C6.5.iii. *Long TR Regime*

If instead of reducing TR, several differently encoded k-lines are acquired following one rf excitation, a very different pulse sequence acceleration method results. The pulse sequences using this method are termed hybrid because, for each excitation, the signal is read using a phase and frequency encoding mixture by the so-called hybrid readout, also known as the echo-train. Hybrid pulse sequences are grouped in families, which are defined according to the type of signals (field echoes versus or spin-echoes), acquired during the hybrid readout: only field-echoes in echo-planar imaging (EPI), only spin-echoes in hybrid-RARE, and a combination of field- and spin-echoes in GraSE.

Hybrid-RARE is the generic denomination for a family of hybrid spin-echo pulse sequences, which includes the original single shot RARE sequence, as well as the more modern variants termed fast spin-echo (FSE) and turbo spin-echo (TSE). The defining architecture of these pulse sequences (Fig. C-6) is the use of one 90° excitation radiofrequency pulse followed by the so-called hybrid-RARE readout, which generates and acquires multiple spin-echo signals with different phase encodings. Accordingly, the hybrid-RARE readout can be viewed schematically as a succession of pulse sequence elements modules (Fig. C-6) of equal duration, which acquire differently phase encoded spin-echo k-lines. Hybrid-RARE pulse sequences are accelerated versions of the CSE pulse sequences.

The pulse sequence variables associated with the hybrid-RARE readout are: 1) the number of k-lines acquired by the hybrid-RARE readout, which is known as echo-train-length (ETL) or turbo factor, depending on the manufacturer. 2) The time interval elapsed between consecutive spin-echo signals, known as echo spacing (ES), and 3) the effective-echo time ($TE_{eff}$) which determines the extent of $T_2$ weighting in a form analogous to the variable TE in CSE sequences. $TE_{eff}$ is defined as the time of acquisition of the k-line(s) with zero or near zero phase encoding which are positioned centrally in k-space. Hybrid-RARE sequences are faster because they acquire imaging data at times during which their CSE counterparts are inactive.

In a form very analogous to CSE sequences, the parameters TR, $TE_{eff}$, and flip angle can be varied to manipulate the contrast weighting of hybrid-RARE pulse sequences. Since the central k-lines are dephased the least by the phase encoding gradients, they contain the strongest MR signals. For this reason the time of acquisition, $TE_{eff}$, of the center of k-space determines the overall level of $T_2$ weighting of the image. The outer portions of k-space contain the weaker MR signals that contribute primarily to image detail and consequently, the times of acquisition of these k-lines determine the sharpness of the image.

Instrument manufacturers differ in their implementations of hybrid-RARE with regard to the scheme used to assign the order in which individual k-lines are acquired (profile order). The profile order specifies the mechanics for varying $TE_{eff}$ and determines the flexibility in the level of $T_2$-weighting of hybrid-RARE scans. The exact forms in which profile the different manufacturers implement orders can be complex and is beyond the scope of this introduction to the subject. However, in the interest of completeness, two very general profile orders, one suitable for short $TE_{eff}$ applications and one for long $TE_{eff}$ applications, are delineated in the following:

Short $TE_{eff}$, which are necessary for proton density-weighted and $T_1$-weighted scans, can be achieved by positioning the k-lines at the center

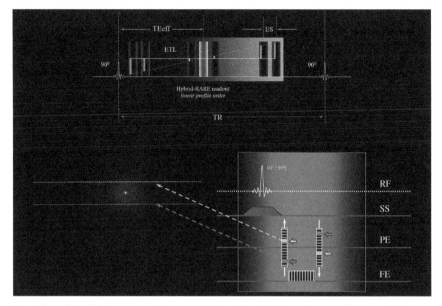

Fig. C-9. Simplified timing diagram of the hybrid-RARE (a.k.a. FSE or TSE) pulse sequence (top). After the excitation pulse, a train of ETL spin-echoes played out. Each spin-echo signal is phase encoded differently (bottom right). In this way, each spin-echo signal is used to form a different line in k-space (bottom left) thus accelerating the acquisition relative to conventional spin-echo, which would have an ETL of one.

of k-space at the beginning of the hybrid-RARE readout. The other k-lines at increasing separation from the center of k-space may be acquired later in the train in an alternating pattern, leading to a profile order referred generically as centric.

The long $TE_{eff}$ that are necessary for $T_2$-weighted scans are commonly achieved by positioning the k-lines at the center of k-space at the midpoint of a long the hybrid-RARE readout. The other k-lines are positioned symmetrically in the hybrid-RARE readout and acquired in a linear pattern, thus leading to a profile order referred as linear.

### C6.6. *Modular Description of Pulse Sequences*

The primary tasks that any MRI pulse sequence must perform are: 1) excitation, 2) spatial encoding, and 3) signal readout. With such basic MRI pulse sequences, contrast weighting can be manipulated by lengthening TE to increase $T_2$ ($T_2^*$)-weighting, or by shortening TR to increase $T_1$-weighting. Pure proton density weighting is obtained in the limiting case TR $\longrightarrow$ $\infty$ and TE $\longrightarrow$ 0, whereby the residual $T_1$- and $T_2$-weightings are vanishing small.

Imparting contrast weighting to qMRI parameters other than $T_1$ and $T_2(T_2^*)$ is achieved by means of additional gradient and rf pulses, in the form either pre-excitation pulse modules that affect primarily the longitudinal magnetization available for excitation, or by means of post-excitation pulse modules that affect primarily the complex transverse magnetization. Pre-excitation pulses are useful for quantifying the parameters of the semisolid pool *via* MT preparation, the longitudinal and transverse relaxation times in the rotating frame *via* spin locking pulses, $T_1$ *via* inversion recovery, and the fat content by any of several spectrally selective methods. As studied above, the post-excitation gradient pulses are particularly useful for quantifying the kinetic qMRI parameters *via* balanced gradient pulse pairs.

A general MRI pulse sequence may be described as consisting of four separate modules as illustrated in schematic form in **Fig.** C-10: 1) pre-excitation weighting, 2) excitation, 3) post-excitation weighting, and 4) signal readout. As reviewed above, several schemes for encoding and signal readout are possible: from conventional readouts reading one spatially encoded signal per TR (gradient-echo, spin-echo, stimulated echo) to hybrid readouts including FSE, GraSE, and EPI.

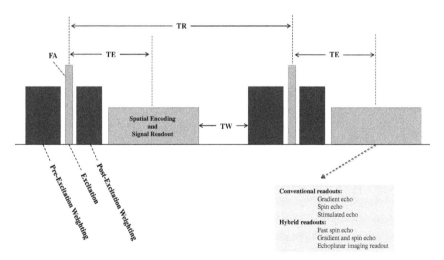

Fig. C-10. Modular representation of a general MRI pulse sequence. Contrast weighting can be obtained and manipulated by either varying the main pulse sequence variables (TR, TE, FA) of by applying pulses before or after magnetization excitation.

C6.6.i. *Pre-Excitation Modules*

Several possibilities are available for the spectral fat suppression pulse module. These include the original chemical shift selective (CHESS) technique (A Haase, Frahm, Hanicke, & Matthaei, 1985), which uses a narrow bandwidth chemically selective 90° pulse with sinc shaped envelope and with carrier frequency centered about the main fat peak. This rf pulse is followed by strong crushers gradients to strongly dephase the recently formed transverse magnetization of the fat and immediately thereafter, the actual imaging pulse sequence begins; since the fat longitudinal magnetization is null, only the water component contributes to the imaging signals. A close relative of CHESS is the method of spectral presaturation with inversion recovery (SPIR) whereby a slightly larger flip angle –typically 110°-120°-- is used to tip the longitudinal magnetization to a negative value. As with CHESS, the generated transverse magnetization is dephased using a crusher gradient pulse, and subsequently, the imaging pulse sequence is started at the null point of the longitudinal magnetization recovery. In this manner, SPIR combines the benefits of CHESS with some of the beneficial fat suppression effects of short tau inversion recovery. Fat suppression quality with CHESS and SPIR is sensitive to presence of $B_0$ and $B_1$ inhomogeneities and these adverse effects increase as function of increasing $B_0$.

For high $B_0$ applications (> 3.0T), use of more robust rf pulses such as the so-called adiabatic pulses becomes necessary. Adiabatic pulses are designed according to an analytical solution of the Bloch equation (Rosenfeld, *et al.*, 1997) using as a driving function a complex hyperbolic secant pulse whereby the pulse amplitude and frequency modulation are functions of time.

$$\left. \begin{array}{l} B_1(t) = B_{1\,max}\sec h(\beta t) \\ \Delta\omega(t) = -\alpha\beta\tanh(\beta t) \end{array} \right\} \quad \text{for} -\frac{T}{2} \le t \le \frac{T}{2} \quad \text{(Eq. C6.6–1)}$$

where the $B_{1\,max}$ is the maximum amplitude of the pulse (typically a few $\mu$T), $\beta$ is the modulation angular frequency, T is the pulse duration, and $\alpha$ is a bandwidth-determining dimensionless parameter. As shown by Silver *et al.* (Silver, *et al.*, 1985), the analytical Bloch equation solution is such that under the appropriate conditions, use of the sech/tanh adiabatic pulse above results in chemically selective magnetization inversion of very high spatial homogeneity. This is because, above a certain threshold, the flip angle becomes independent of the pulse amplitude. This is a unique mathematical feature of adiabatic pulses; hence the importance of adiabatic

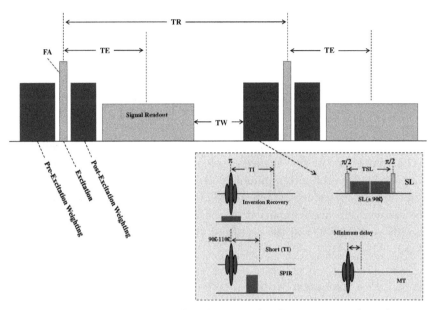

Fig. C-11.   Pre-excitation modules for achieving either $T_1$ contrast weighting by inversion recovery, or $T_{1\rho}$ contrast weighting with a spin-locking pulse module, magnetization transfer contrast with an MT prepulse or chemical selectivity (fat or water suppression).

pulses for high field applications (Balchandani & Spielman, 2008). In particular, such an inversion pulse can be used for fat suppression in a technique referred to as spectral attenuated inversion recovery (SPAIR), which is a chemically selective inversion recovery experiment.

The IR pulse module consists of a slice selective 180° pulse with a sinc waveform or with an adiabatic waveform for high field applications (sees discussion above).

For the spin-lock preparation pulse module (**Fig.** C-11), a nonselective $\pi/2$ pulse excites spins that are then spin-locked in the transverse plane by the application of two phase-alternating SL pulses (90° phase-shifted from the phase of the first $\pi/2$ pulse). Investigators at the University of Pennsylvania (Wheaton, Borthakur, Charagundla, & Reddy, 2004) have found that phase-alternating spin-locking (SL) pulses reduce image artifacts in $T_{1\rho}$ MRI resulting from $B_1$ inhomogeneity. The duration of the spin-locking pulses is denoted as TSL. These investigators used a delay of $20\,\mu\text{sec}$ between SL pulse segments to accommodate for electronic dead time between rf instructions. The second nonselective $\pi/2$ pulse restores the spin-locked magnetization to the longitudinal axis. Finally, a high amplitude gradient (not shown) is applied for destroying residual transverse magnetization. The "$T_{1\rho}$-prepared" longitudinal magnetization at the end of the crusher gradient is described by the equation (G. E. Santyr, Henkelman, & Bronskill, 1989):

$$M_z^{(A)}(\text{TSL}) = M_z^{(\text{eq-A})} \exp\left(-\frac{\text{TSL}}{T_{1\rho}^{(A)}}\right) \qquad \text{(Eq. C6.6–2)}$$

The MT pulse preparation module consists of a single rf pulse typically with a Gaussian-shape or sinc-shaped envelope and the imaging pulse sequence is applied thereafter with minimum delay. A frequency offset, a bandwidth, and the peak $B_1$ value characterize MT prepulses; some researchers report the nominal flip angle imparted by the MT pulse. A typical MT protocol (Smith *et al.*, 2009) will use a pulse duration of 10–15 ms and peak $B_1$ of 15 μT.

### C6.6.ii. *Post-Excitation Modules*

Post-excitation pulse sequence modules are primarily for sensitizing to kinetic qMRI parameters; these include modules for diffusion-perfusion sensitization and modules for velocity encoding (see insert in **Fig. C-12**).

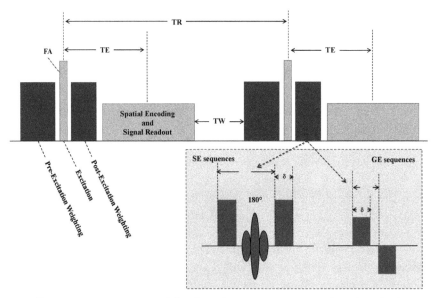

Fig. C-12. Post-excitation modules for achieving contrast weighting in the kinetic variables of diffusion, perfusion, or flow and displacement.

In both cases, the main idea is to apply balanced gradient pulses whereby the transverse magnetization of immobile spins is unaffected while for spins in motion, the transverse magnetization is not rephased. Diffusion-perfusion weighted MRI is done mostly with spin-echo-like pulse sequences and therefore, two gradient pulses of the same polarity are applied one before and one after the 180° refocusing pulse and these can be applied in any direction. On the other hand, velocity qMRI is done mostly with gradient-echo pulse sequences and the balanced velocity weighting gradient pulses are of opposite polarity; again, velocity sensitization can be accomplished in any spatial direction.

## References

Balchandani, P., & Spielman, D. (2008). Fat suppression for 1H MRSI at 7T using spectrally-selective adiabatic inversion recovery. *Magnetic resonance in medicine, 59*(5), 980.

Basser, P. J., Mattiello, J., & LeBihan, D. (1994). Estimation of the effective self-diffusion tensor from the NMR spin echo. *J Magn Reson B, 103*(3), 247–254.

Bloch, F. (1946). Nuclear Induction. *Phys Rev, 70*, 460.

Chaplin, M. (2010). Water Structure and Science (http://www1.lsbu.ac.uk/water/).

Deoni, S. C., Rutt, B. K., & Peters, T. M. (2003). Rapid combined T1 and T2 mapping using gradient recalled acquisition in the steady state. *Magn Reson Med, 49*(3), 515–526.

Dixon, W., Engels, H., Castillo, M., & Sardashti, M. (1990). Incidental magnetization transfer contrast in standard multislice imaging. *Magnetic resonance imaging, 8*(4), 417–422.

Ernst, R., Bodenhausen, G., & Wokaun, A. (1990). *Principles of nuclear magnetic resonance in one and two dimensions*: Oxford University Press, USA.

Haacke, E., Mittal, S., Wu, Z., Neelavalli, J., & Cheng, Y. (2009). Susceptibility-weighted imaging: technical aspects and clinical applications, part 1. *American Journal of Neuroradiology, 30*(1), 19.

Haase, A. (1990). Snapshot FLASH MRI. Applications to T1, T2, and chemical-shift imaging. *Magn Reson Med, 13*(1), 77–89.

Haase, A., Frahm, J., Hanicke, W., & Matthaei, D. (1985). 1H NMR chemical shift selective (CHESS) imaging. *Physics in Medicine and Biology, 30*, 341–344.

Hahn, E. (1950). Spin Echoes. *Physical Review, 80*, 580–594.

Henkelman, R. M., Huang, X., Xiang, Q. S., Stanisz, G. J., Swanson, S. D., & Bronskill, M. J. (1993). Quantitative interpretation of magnetization transfer. *Magn Reson Med, 29*(6), 759–766.

Ibrahim, T., Lee, R., Abduljalil, A., Baertlein, B., & Robitaille, P. (2001). Dielectric resonances and B1 field inhomogeneity in UHFMRI: computational analysis and experimental findings. *Magnetic resonance imaging, 19*(2), 219–226.

Joseph, P. M., Axel, L., & O'Donnell, M. (1984). Potential problems with selective pulses in NMR imaging systems. *Med Phys, 11*(6), 772–777.

Melki, P. S., & Mulkern, R. V. (1992). Magnetization transfer effects in multislice RARE sequences. *Magn Reson Med, 24*(1), 189–195.

Moran, P., & Hamilton, C. (1995). Near-resonance spin-lock contrast. *Magnetic resonance imaging, 13*(6), 837–846.

Morrison, C., Stanisz, G., & Henkelman, R. M. (1995). Modeling magnetization transfer for biological-like systems using a semi-solid pool with a super-Lorentzian lineshape and dipolar reservoir. *J Magn Reson B, 108*(2), 103–113.

Oppelt, A., Graumann, R., Barfuss, H., Fischer, H., Hartl, W., & Schajor, W. (1986). FISP: a new fast MRI sequence. *Electromedica, 54*(1), 15–18.

Parker, D. L., Du, Y. P., & Davis, W. L. (1995). The voxel sensitivity function in Fourier transform imaging: applications to magnetic resonance angiography. *Magn Reson Med, 33*(2), 156–162.

Perkins, T., & Wehrli, F. (1986). CSF signal enhancement in short TR gradient echo images. *Magnetic resonance imaging, 4*(6), 465–467.

Rosenfeld, D., Panfil, S., & Zur, Y. (1997). Design of adiabatic pulses for fat-suppression using analytic solutions of the Bloch equation. *Magnetic Resonance in Medicine, 37*(5), 793–801.

Santyr, G. (1993). Magnetization transfer effects in multislice MR imaging. *Magnetic resonance imaging, 11*(4), 521–532.

Santyr, G. E., Henkelman, R. M., & Bronskill, M. J. (1989). Spin locking for magnetic resonance imaging with application to human breast. *Magn Reson Med, 12*(1), 25–37.

Scheffler, K., & Hennig, J. (2001). T(1) quantification with inversion recovery TrueFISP. *Magn Reson Med, 45*(4), 720–723.

Scheffler, K., & Lehnhardt, S. (2003). Principles and applications of balanced SSFP techniques. *European radiology, 13*(11), 2409–2418.

Shankar, R. (2007). *Principles of quantum mechanics.* New York, USA: Springer.

Silver, M., Joseph, R., & Hoult, D. (1985). Selective spin inversion in nuclear magnetic resonance and coherent optics through an exact solution of the Bloch-Riccati equation. *Physical Review A, 31*(4), 2753–2755.

Smith, S., Golay, X., Fatemi, A., Mahmood, A., Raymond, G., Moser, H., *et al.* (2009). Quantitative magnetization transfer characteristics of the human cervical spinal cord in vivo: application to adrenomyeloneuropathy. *Magnetic resonance in medicine, 61*(1), 22.

Stejskal, E. O. (1965). Use of Spin Echoes in a Pulsed Magnetic-Field Gradient to Study Anisotropic Restricted Diffusion and Flow. *Journal of Chemical Physics, Vol. 43*, 3597–3603.

Stejskal, E. O., & Tanner, J. E. (1965). Spin Diffusion Measurements: Spin Echoes in the Presence of a Time-Dependent Field Gradient. *Journal of Chemical Physics, Vol. 42*, 288–292.

Taheri, S., & Sood, R. (2006). Spin-lock MRI with amplitude-and phase-modulated adiabatic waveforms: an MR simulation study. *Magnetic resonance imaging, 24*(1), 51–59.

Torrey, H. C. (1956). Bloch Equations with Diffusion Terms. *Physical Review,* vol. *104*(3), 563–565.

Turner, R., Le Bihan, D., & Chesnick, A. S. (1991). Echo-planar imaging of diffusion and perfusion. *Magn Reson Med, 19*(2), 247–253.

Turner, R., Le Bihan, D., Maier, J., Vavrek, R., Hedges, L. K., & Pekar, J. (1990). Echo-planar imaging of intravoxel incoherent motion. *Radiology, 177*(2), 407–414.

Wheaton, A. J., Borthakur, A., Charagundla, S. R., & Reddy, R. (2004). Pulse sequence for multislice T1rho-weighted MRI. *Magn Reson Med, 51*(2), 362–369.

# D. ELEMENTS OF RELAXATION THEORY

## D1. Introduction

The trajectory of the magnetization vector during NMR excitation is illustrated in Fig. D-1. The tip of the magnetization vector moves along a spiral curve on the surface of a sphere. After excitation, the magnetization has been tipped away from equilibrium and the phenomenon of relaxation takes over. Although the magnetizations of all substances behave almost identically during excitation, the magnetization trajectories of the relaxing tissues can differ drastically from one another depending primarily on the relaxation times, as illustrated by computer simulation in Fig. D-2. For substances with $T_2 \sim T_1$, the tip of the magnetization vector would move on the surface of a cone; such materials include certain paramagnetic solutions (N Bloembergen, 1957). Relaxation of pure water ($T_2 \sim 0.5T_1$) is similar but in this case the cone is exponentially deformed about the z-axis. Such deformations increase for typical soft tissues whereby $T_2 \sim 0.1T_1$ (see Fig. D-2).

The phenomenon of NMR relaxation can be described at three increasingly fundamental levels of physical significance and theoretical complexity: 1) the phenomenological level, in which magnetization dynamics is described with the Bloch equations or its generalized variants and the relaxation times are treated as phenomenological parameters to be derived experimentally. 2) The semi-classical level, in which magnetization dynamics is described with the (generalized-) Bloch equations and the relaxation times are calculate *via* spin packet statistical quantum mechanics, and finally, 3) the full quantum mechanical level, in which the magnetization dynamics and the interactions with surrounding are treated quantum mechanically.

This chapter is concerned with the semi-classical description of the NMR relaxation phenomenon with focus on the primary relaxation times

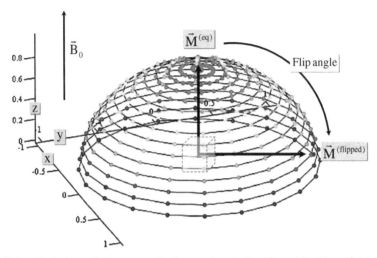

Fig. D-1. Trajectory of the magnetization vector during the application of a brief rf pulse. The tip of the vector moves along a spiral curve on the surface of a sphere. Simulation assumes relaxation effects are negligible ($T_1$, $T_2 \gg$ rf pulse duration).

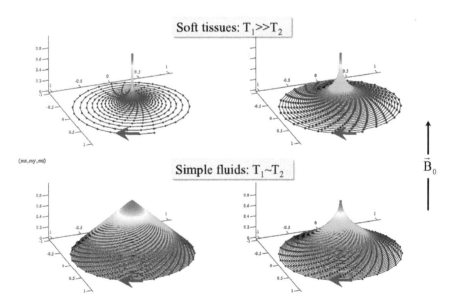

Fig. D-2. Patterns of relaxation for different $T_1$ and $T_2$ combinations (see text).

$T_1$ and $T_2$, and the theoretical insights resulting from comparing the multi-particle quantum mechanical equations of motion *vs.* the Bloch equations.

## D1.1. *Relaxation Fundamentals*

NMR relaxation encompasses all physical processes by which spin packets return to the state of thermal equilibrium after being perturbed, for example by applying an excitatory rf pulse. Relaxation processes are customarily divided into two types: spin-lattice relaxation and spin-spin relaxation. Spin-lattice relaxation refers to the equilibration of spin populations back to the Boltzmann distribution values. Spin-spin relaxation on the other hand, refers to the decay of transverse coherences back to the fully randomized null value. Although, the recovery of longitudinal magnetization is purely caused by spin-lattice relaxation mechanisms, the decay of transverse magnetization is caused by both, spin-spin relaxation mechanisms, giving rise to the secular $T_2$ relaxation time constant, and by spin-lattice relaxation giving rise to the non-secular $T_2$ relaxation time.

The relaxation rates, *i.e.* the inverse of the relaxation times, constitute measures of the relaxation effectiveness of the different interaction mechanisms present in tissue. Specifically, the longitudinal recovery or spin-lattice relaxation rate provides a combined measure of the relaxation effectiveness of the following three families of processes:

1) the $^1$H-proton-$^1$H-proton dipole-dipole interactions that are modulated by tissue micro-kinetic motions, specifically by molecular rotations and translations.
2) Relaxation processes caused by interactions between $^1$H-protons with the unpaired electrons of paramagnetic solutes, such as naturally dissolved oxygen and possibly exogenous contrast agents.
3) Magnetization exchange processes with $^1$H-protons of the semisolid component (B).

Hence, the total longitudinal relaxation rate of the liquid pool in tissue can be written as the sum of three terms, specifically:

$$\frac{1}{T_1^{(A)}} = \frac{1}{T_1^{(A,kin)}} + \frac{1}{T_1^{(A,para)}} + \frac{1}{T_1^{(A,ex)}} \qquad \text{(Eq. D1.1-1)}$$

where the kinetic term applies to the mobile $^1$H-protons of the liquid pool, the paramagnetic term reflects the relaxation processes involving unpaired electrons of paramagnetic solutes, and the exchange term reflects

$T_1$ relaxation contributions caused by spin exchange between the liquid and the semisolid pool.

Physical processes that contribute to the decay of the transverse magnetization include energy-conserving spin-spin processes that reduce the level of precession coherence among spins: this is known as the secular component of the $T_2$ rate. In addition, the $T_1$ relaxation processes mentioned above lead to the so-called non-secular component of the $T_2$ relaxation rate, which is also known as the $T_1$ contribution to $T_2$ (Thomas, 1986). In summary, the transverse relaxation rate is the sum of two terms:

$$\frac{1}{T_2^{(A)}} = \frac{1}{T_2^{(A,sec)}} + \frac{1}{2T_1^{(A)}} \qquad \text{(Eq. D1.1–2)}$$

We note that the non-secular contribution to the $T_2$ relaxation rate is equal to only one half of the $T_1$ relaxation rate because each relaxation process reduces the number of spins in transverse state by one, while the same relaxation process would reduce the net number of longitudinally polarized spins by two. Consequently, $T_1$ relaxation processes are half as effective in reducing the transverse magnetization relative to the longitudinal magnetization. Hence, from a pure theoretical standpoint, the following inequality holds: $T_2^{(A)} \leq 2T_1^{(A)}$. In practice however the observed inequality is more strict, specifically: $T_2^{(A)} \leq T_1^{(A)}$.

The effectiveness of the microscopic relaxation mechanisms discussed above increases for processes mediated by oscillatory magnetic fields at the resonance frequency and therefore the relaxation times are functions of the polarizing field strength.

### D1.2. *Rate Equation Analysis: Two-Level Systems*

The processes of resonant-absorption and stimulated-emission cause transitions between the spin-up and spin-down states and are represented by transition rates, specifically $R^{(ab)}$ for absorption and $R^{(em)}$ for stimulated emission. Accordingly, in the absence of any applied external stimuli such as resonant rf fields, the temporal evolutions of the spin-down and spin-up populations satisfy the following system of rate equations:

$$\frac{\partial}{\partial t} n_{up} = -R^{(ab)} n_{up} + R^{(em)} n_{down} \qquad \text{(Eq. D1.2–1)}$$

and

$$\frac{\partial}{\partial t} n_{down} = -R^{(em)} n_{down} + R^{(ab)} n_{up} \qquad \text{(Eq. D1.2–2)}$$

Defining $\Delta n \equiv n_{up} - n_{down}$ and $n \equiv n_{up} + n_{down}$ we can transform the rate equations into:

$$\frac{\partial}{\partial t}\Delta n = -R^{(ab)}(n + \Delta n) + R^{(em)}(n - \Delta n) \qquad \text{(Eq. D1.2–3)}$$

In thermal equilibrium, the population difference is constant, hence:

$$\frac{\partial}{\partial t}\Delta n = 0 \qquad \text{(Eq. D1.2–4)}$$

The above condition implies that the spin-up minus spin-down population difference in thermal equilibrium is a fraction of the total spin density and given by:

$$\Delta n^{(eq)} = n\frac{R^{(em)} - R^{(ab)}}{R^{(em)} + R^{(ab)}} \qquad \text{(Eq. D1.2–5)}$$

We can rewrite

$$\frac{\partial}{\partial t}\Delta n = (R^{(em)} + R^{(ab)})(\Delta n^{(eq)} - \Delta n) \qquad \text{(Eq. D1.2–6)}$$

Furthermore, the longitudinal magnetization is proportional to the population difference at all times, therefore

$$\frac{\partial M_z^{(A)}}{\partial t} = \frac{(M_z^{(eq\text{-}A)} - M_z^{(A)})}{T_1^{(A)}} \qquad \text{(Eq. D1.2–7)}$$

Where, the longitudinal relaxation time $T_1$ has been defined as the inverse of the sum up and down transition rates, *i.e.*:

$$\frac{1}{T_1^{(A)}} = R^{(em)} + R^{(ab)} \qquad \text{(Eq. D1.2–8)}$$

In conclusion, the relaxation rate of the longitudinal magnetization is the sum of the resonant-absorption and stimulated-emission rates. Moreover, the rate equations predict the longitudinal magnetization recovery term of the Bloch equation, which leads to monoexponential recovery under the assumption of constant stimulated emission and absorption rates.

## D2. Spin Packet Quantum Mechanics

### D2.1. *Spin Packet Hamiltonian*

The multi-particle Schrödinger equation of a spin packet reads:

$$i\hbar\frac{\partial}{\partial t}|\psi(t)\rangle = H^{packet}|\psi(t)\rangle \qquad \text{(Eq. D2.1–1)}$$

The packet state is the product of the single particle spin states, specifically:

$$|\psi(t)\rangle = \prod_{\vec{\mu}^{(m)} \subseteq \delta V^{(\text{packet})}(\vec{x})} |\psi_m(t)\rangle \qquad \text{(Eq. D2.1–2)}$$

The Hamiltonian of a spin packet as a whole is the sum of all single-spin Zeeman terms representing the magnetic coupling of each and all spins to the polarizing magnetic field and includes terms that account for the magnetic interactions of each spin with other particles of the packet.

1. Single-spin Zeeman energies
2. Dipole-dipole interactions: resonant absorption and stimulated emission
3. Electron shielding: chemical shift interactions
4. J-coupling

The sum of all single-spin energy operators is the Zeeman Hamiltonian, accordingly we write:

$$\mathbf{H}^{\text{Zeeman}} = - \sum_{\vec{\mu}^{(m)} \subseteq \delta V^{(\text{packet})}(\vec{x})} \mu_{\mathbf{z}}^{(\mathbf{m})}(t) \cdot \vec{B}_0 \qquad \text{(Eq. D2.1–3)}$$

This accounts for the strongest interaction of spins in NMR experimentation, which is the coupling of magnetic moments with the externally applied magnetic field, which is static.

Next in strength, are the direct dipole-dipole couplings, which refer to direct interactions between nuclear spins, as opposed to indirect spin interactions that are mediated by electrons; accordingly the dipolar Hamiltonian is the sum over all *i.e.* intramolecular and intermolecular spin pairs in the packet of the dipole-dipole interaction energies, specifically

$$\mathbf{H}^{\text{D}} = \frac{\mu_0}{2(2\pi)^3} \sum_{\vec{\mu}^{(m)} \subseteq \delta V^{(\text{packet})}(\vec{x})} \vec{\boldsymbol{\mu}}^{(\mathbf{m})}(t)$$

$$\cdot \sum_{n<m} \left\{ \frac{\vec{\boldsymbol{\mu}}^{(\mathbf{n})}(t)}{r_{mn}^3} - \frac{3\vec{r}_{mn}(\vec{\boldsymbol{\mu}}^{(\mathbf{n})}(t) \cdot \vec{r}_{mn})}{r_{mn}^5} \right\} \qquad \text{(Eq. D2.1–4)}$$

The m-n inter-nuclear separation vectors $\vec{r}_{mn}$ are functions of time; in liquids and tissue these inter-nuclear separations vectors change randomly both in orientation and in magnitude as functions of time because of molecular rotational and translational motions, which are of thermal origin. For the two $^1$H-protons of a single water molecule, the inter-nuclear separation is approximately constant. In an isotropic liquid, the rotational motions of molecules leads to motion averaging of the spin interactions such that all intramolecular spin interactions are averaged to the isotropic value.

The Chemical Shift Hamiltonian represents the shielding effect that orbital electrons have on the magnetic field at the position of a $^1$H-proton. We are concerned in this book only with the isotropic case whereby, the chemical shift phenomenon can be described by the simple Hamiltonian:

$$\mathbf{H}^{CS} = -\sum_{\vec{\mu}^{(m)} \subseteq \delta V^{(packet)}(\vec{x})} \mu_z^{(m)}(t) \cdot \sigma_m \vec{B}_0 \qquad \text{(Eq. D2.1–5)}$$

in which, $\sigma_m$ is a dimensionless scalar known as the isotropic chemical shift of a given $^1$H-proton type and is typically only a few parts per million.

We finish this section with a brief discussion of the J-Coupling Hamiltonian, specifically a type of indirect spin-spin interaction that is mediated by molecular bonding electrons. This coupling arises from the interaction of different spin states through the chemical bonds in a molecule and results in the splitting of NMR spectral features. It is given by:

$$\mathbf{H}^{JC} = 2\pi\gamma^{-2} \sum_{(\vec{\mu}^{(m)}, \vec{\mu}^{(n)}) \subseteq \delta V^{(packet)}(\vec{x})} \vec{\mu}^{(m)}(t) \cdot \mathbf{J}_{m,n} \cdot \vec{\mu}^{(n)}(t)$$

$$\text{(Eq. D2.1–6)}$$

where $\mathbf{J}_{m,n}$ is the J-coupling tensor, a $3 \times 3$ real matrix.

Chemical shift and J-coupling represent two fundamental links between NMR spectroscopy and chemistry by providing analytical tools for unraveling chemical structure from NMR spectra. In the context of qMRI however, these two phenomena are of importance primarily for the analysis and separation of adipose tissues from aqueous tissues, as well as for interpreting the abnormally elongated $T_2$ decay of fat as observed with turbo spin-echo sequences. These are highly rf-refocused pulse sequences thus capable of competing with J-coupling decay mechanisms.

## D2.2. *Random Field Relaxation: Local Field Equations*

We model a spin packet as a collection of solitary spins each of which is subjected to two superposed magnetic fields. First, the polarizing main magnetic field, which is static and has the same value for all spins of the same chemical type *i.e.* water *vs.* lipid protons. Second, a much weaker and time varying magnetic field, we refer to this as the local magnetic field. We assume that these solitary spins are independent of each other and that the description of their interactions with the environment is represented in full by the randomly fluctuating local magnetic fields. We further assume that the local fields of any two different spins are uncorrelated to each other. However, each local magnetic field is time correlated to itself with a certain correlation time, which depends on the motional degrees of freedom of molecules in the medium. We can therefore replace the multi-particle Schrödinger equation above (see Eq. D2.1–1) with a system of uncoupled single-particle Schrodinger equations of the form:

$$i\hbar\frac{\partial}{\partial t}|\psi_m(t)\rangle = -\vec{\mu}(t) \cdot (\vec{B}_0(1 - \sigma_m) + \vec{B}_{loc}^m(t))|\psi_m(t)\rangle$$

(Eq. D2.2–1)

where the local magnetic field is the dipolar field, *i.e.*

$$\vec{B}_{loc}^{(m)}(t) = \frac{\mu_0}{4\pi}\sum_{n<m}\left\{\frac{3\vec{r}_{mn}(t)(\vec{\mu}^{(n)}(t)\cdot\vec{r}_{mn}(t))}{r_{mn}^5(t)} - \frac{\vec{\mu}^{(n)}(t)}{r_{mn}^3(t)}\right\}$$

(Eq. D2.2–2)

is given in terms of magnetic moment expectation values:

$$\vec{\mu}^{(n)}(t) = \langle\psi_n(t)|\vec{\boldsymbol{\mu}}^{(n)}|\psi(t)\rangle$$

This problem is special case of the general problem studied earlier (see Section C2.3) of a solitary spin in a time dependent magnetic field. Hence, the same equations of motion for the expectation values of each magnetic moment apply, specifically:

$$\frac{d\mu_T^{(m)}}{dt} = -i2\pi\gamma[(1 - \sigma_m)B_0 + B_z^{(m)}(t)]\mu_T^{(m)}(t) + i2\pi\gamma B_T^{(m)}(t)\mu_z^{(m)}(t)$$

(Eq. D2.2–3)

and

$$\frac{d\mu_z^{(m)}}{dt} = 2\pi\gamma\text{Im}\{B_T^{(m)}(t)^*\mu_T^{(m)}(t)\}$$

(Eq. D2.2–4)

These differential equations are remarkably similar to the transverse and longitudinal Bloch equations; therefore, we will use them for transitioning into the semi-classical description by performing suitable ensemble averages, and in the process derive equations for the relaxation rates.

## D2.3. *The Bloch Equation Connection*

We are interested in magnetic fields that include both a randomly fluctuating local field, which is specific to the m-spin, as well as any potential applied field that acts equally on all spins, hence

$$B_T^{(m)}(t) = B_T^{(App)}(t) + B_T^{(Int)}(m, t) \qquad \text{(Eq. D2.3–1)}$$

and

$$B_z^{(m)}(t) = B_z^{(App)}(t) + B_z^{(Int)}(m, t) \qquad \text{(Eq. D2.3–2)}$$

Accordingly, the single particle magnetic moment expectation value equations (Section C.2) become:

$$\frac{d\mu_T^{(m)}}{dt} = -i\,2\pi\gamma\big[(1 - \sigma_m)B_0 + B_z^{(Int)}(m, t)\big]\mu_T^{(m)}(t)$$
$$+ i2\pi\gamma\big(B_T^{(App)}(t) + B_T^{(Int)}(m, t)\big)\mu_z^{(m)}(t) \qquad \text{(Eq. D2.3–3)}$$

and

$$\frac{d\mu_z^{(m)}}{dt} = 2\pi\gamma\text{Im}\big\{\big(B_T^{(App)}(t) + B_T^{(Int)}(m, t)\big)^*\mu_T^{(m)}(t)\big\}$$
$$\text{(Eq. D2.3–4)}$$

The objective is to perform ensemble averages of these equations to establish a connection with semi-classical physics. By performing ensemble averages on the left hand sides, we find

$$\sum_{\mu_T^{(m)} \subseteq \delta V^{(packet)}(\vec{x})} \frac{d\mu_T^{(m)}}{dt} = \frac{dm}{dt} \qquad \text{(Eq. D2.3–5)}$$

and,

$$\sum_{\mu_T^{(m)} \subseteq \delta V^{(packet)}(\vec{x})} \frac{d\mu_z}{dt} = \frac{dM_z}{dt} \qquad \text{(Eq. D2.3–6)}$$

which therefore allows direct comparisons of the ensemble averages of the right hand side terms with those of the Bloch equations:

$$\frac{dm}{dt} = -i2\pi\gamma B_z^{(Exp)}m - \frac{m}{T_2^{(A,sec)}} - \frac{m}{2T_1^{(A)}} + i2\pi\gamma B_{1T}^{(Exp)}M_z$$
$$\text{(Eq. D2.3–7)}$$

and,

$$\frac{dM_z}{dt} = -2\pi\gamma \, \text{Im}\{B_{1T}^{(Exp)*}m\} + \frac{\left(M_z^{(eq)} - M_z\right)}{T_1^{(A)}} \qquad (Eq. \ D2.3\text{--}8)$$

Of interest here are the following resulting three equalities, which result from comparing Eq. D2.3–7, 8 *vs.* Eq. D2.3–3, 4, specifically:

$$\frac{1}{T_2^{(A,sec)}(\vec{x})} = \frac{i2\pi\gamma}{m(\vec{x},t)} \sum_{\mu_T^{(m)} \subseteq \delta V^{(packet)}(\vec{x})} B_z^{(Int)}(m,t)\mu_T^{(m)}(t)$$

$$(Eq. \ D2.3\text{--}9)$$

and

$$\frac{1}{T_1^{(A)}(\vec{x})} = -\frac{i4\pi\gamma}{m(\vec{x},t)} \sum_{\mu_T^{(m)} \subseteq \delta V^{(packet)}(\vec{x})} B_T^{(Int)}(m,t)\mu_z^{(m)}(t)$$

$$(Eq. \ D2.3\text{--}10)$$

and

$$\frac{1}{T_1^{(A)}(\vec{x})} = \frac{2\pi\gamma}{\left(M_z^{(eq)}(\vec{x}) - M_z(\vec{x},t)\right)}$$

$$\times \text{Im}\left\{ \sum_{\mu_T^{(m)} \subseteq \delta V^{(packet)}(\vec{x})} B_T^{(Int)}(m,t)^* \mu_T^{(m)}(t) \right\}$$

$$(Eq. \ D2.3\text{--}11)$$

These results show that: 1) secular-$T_2$ relaxation is caused by the z-component of the randomly fluctuating field, and 2) $T_1$ relaxation is caused by the transverse components of the randomly fluctuating field. In other words, secular-$T_2$ relaxation is a loss of coherence phenomenon that is secondary to dephasing caused by random fluctuations of the Larmor frequency, and $T_1$ relaxation is caused by resonant absorption and stimulated emission processes.

## D2.4. *Random Field Relaxation*

For any given spin of the packet, each x-, y-, z-component of the random field switches randomly and rapidly from positive to negative values, thus resulting in a null temporal average. Accordingly, in the simplest model, one assumes that the fluctuating fields have zero temporal averages over sufficiently long periods, *i.e.*:

$$\left\langle B_x^{(Int)}(m,t) \right\rangle = \left\langle B_y^{(Int)}(m,t) \right\rangle = \left\langle B_z^{(Int)}(m,t) \right\rangle = 0$$

(Eq. D2.4–1)

Moreover, the ergodic hypothesis, which says that all accessible microstates of a system with a large number of particles are equiprobable over a long period of time, is often invoked for treating equivalently and interchangeably temporal and ensemble averages. Hence, in the case of the random fields responsible for NMR relaxation, the spin packet averages are also zero at any time,

$$\left\langle B_{x,y,z}^{(Int)}(m,t) \right\rangle = \sum_{\vec{\mu}^{(m)} \subseteq \delta V^{(packet)}(\vec{x})} B_{x,y,z}^{(Int)}(m,t) = 0 \qquad \text{(Eq. D2.4–2)}$$

Since the averages are null, one therefore needs a different way of quantifying the magnitudes of the fluctuating field components. This is done *via* the squares of the magnetic fields,

$$\left\langle \left( B_{x,y,z}^{(Int)}(m,t) \right)^2 \right\rangle \neq 0 \qquad \text{(Eq. D2.4–3)}$$

which are always positive.

Additionally, one needs to quantify how rapidly the random fields fluctuate. This is done with the autocorrelation functions, which are defined as follows:

$$G_{x,y,z}(\tau) \equiv \left\langle B_{x,y,z}^{(Int)}(m,t) B_{x,y,z}^{(Int)}(m,t+\tau) \right\rangle \neq 0 \qquad \text{(Eq. D2.4–4)}$$

Under the so-called stationary assumption, the autocorrelation functions are independent of observation time t and are functions of relative time $\tau$ only. Furthermore, the random statistical nature of the molecular motions, which determine the time dependence of the randomly fluctuating magnetic fields, justifies the assumption that the autocorrelation functions are also spin-independent *i.e.* independent of spin number m and decrease rapidly as function of $\tau$. These functions are often modeled with exponential functions,

$$G_{x,y,z}(\tau) = \hat{B}_{x,y,z}^{(Int)^2} \exp\left( -\frac{|\tau|}{\tau_c} \right) \qquad \text{(Eq. D2.4–5)}$$

where we have assumed isotropy of the random fluctuations. In such model, $\tau_c$ is a time constant known as the correlation time, and $\hat{B}_{x,y,z}^{(Int)}$ is the mean x-, y-, or z-component of the randomly fluctuating field. The correlation time is a qualitative measure of how long it takes the random field to change sign. Rapidly fluctuating fields have short correlation times and conversely, slowly fluctuating fields have long correlation times. The exponential form of this correlation function can be shown (Dixon & Ekstrand, 1982) to follow from the diffusion equation describing molecules that are individually undergoing complex Brownian motions, but the average behavior can be described as a diffusion. Clearly the larger the difference t-t', the less correlation there is between the (random) field between times t and t', thus the smaller the average product in Eq. D2.4–3. The correlation time is a measure of the time over which the intramolecular magnetic fields change appreciably specifically by $1/e$ and for water at room temperature, $\tau_c \sim 6\,\text{psec}$.

### D2.5. *Spectral Density*

The spectral density of the fluctuating field is the Fourier transform of the autocorrelation function, and is a measure of the field per unit of frequency that is available for relaxation.

$$J(\omega) \equiv \int_{-\infty}^{\infty} G(\tau)\exp(-i\omega\tau)d\tau \qquad \text{(Eq. D2.5–1)}$$

For a fluctuating field along the z-axis, with an exponentially decaying autocorrelation function,

$$J_{x,y,z}(\omega) = 2\langle \hat{B}_{x,y,z}^{(Int)^2}\rangle \frac{\tau_c}{1+\omega^2\tau_c^2} \qquad \text{(Eq. D2.5–2)}$$

If the random field fluctuates rapidly, then the spectral density is broad because the correlation time is short. Conversely, if the random field is slowly fluctuating, then the correlation time is long and the spectral density is narrow.

### D2.6. *Relaxation in Liquids: Kinetic Dipolar* $\mathbf{T_1}$ *and* $\mathbf{T_2}$ *Relaxation Mechanisms*

Liquid water is probably the single most studied substance in biology and its MR properties are well described with theoretical models developed more than fifty years ago (N. Bloembergen, Purcell, & Pound, 1948; Carr &

Purcell, 1954; Kubo & Tomita, 1954; Solomon, 1955; Torrey, 1953), which are still current.

In liquids, the major $T_1$ relaxation mechanisms stem from dipolar magnetic interactions with the randomly fluctuating local magnetic field, which is modulated by the motions of the neighboring dipoles; specifically rotations (molecular tumbling) and linear translations (diffusion).

In 1948, Bloembergen, Purcell, and Pound developed a quantum mechanics based relaxation theory (N. Bloembergen, *et al.*, 1948), known as the BPP theory, which successfully explained $T_1$ relaxation of liquids as caused by the fluctuating local dipolar fields. This seminal theoretical work is still current and applies to wide variety of liquids of different viscosities, with some exceptions including ordered liquids such liquid crystals and polymer melts (Callaghan, 1993).

Dipole-dipole interactions can take place between two protons of the same molecule *i.e.* intra-molecular dipole-dipole interactions or between protons of different molecules *i.e.* inter-molecular dipole-dipole interactions. $T_1$-relaxation effects caused by the intra-molecular proton-proton interactions are modulated by the rotational molecular motions or molecular tumbling. On the other hand, the $T_1$-relaxation effects caused by inter-molecular interactions are primarily modulated by translational motions (Torrey, 1953). Accordingly, the total kinetic $T_1$ relaxation rate can be written as:

$$\left[ \frac{1}{T_1^{(A)}} \right]_{(kin)} = \left[ \frac{1}{T_1^{(A)}} \right]_{(rot)} + \left[ \frac{1}{T_1^{(A)}} \right]_{(trans)} \qquad \text{(Eq. D2.6–1)}$$

These two $T_1$ relaxation rate contributions were studied theoretically using quantum mechanics (Solomon, 1955) and random walk theory (Torrey, 1953). Central to these theories is the concept of correlation time, which is a statistical measure of the time interval during which the magnetic field at a proton site remains electrodynamically correlated to its earlier values and does not change abruptly by stochastic interactions, such as molecular collisions. Accordingly, the rotational and translational correlation times are temporal measures for molecular reorientation and translation in tissue, respectively.

Deriving formulas for the rotational and translational $T_1$ relaxation rates in terms of the correlation times requires use of advanced quantum physics tools including time-dependent perturbation theory and the density matrix formalism of quantum statistic. Review of these involved theoretical treatments is beyond the scope of this book and the reader is referred the original papers or any of the authoritative textbooks on this subject matter (Abragam, 1961; Callaghan, 1993; Slichter, 1990). Here, we limit ourselves to reproducing the essential BPP results in the notation of Callaghan (Callaghan, 1993), which are expressed in terms of the spectral density functions $J^{(q)}(\omega)$ (of order q), specifically:

$$\left[\frac{1}{T_1^{(A)}}\right]_{(rot)} = \left(\frac{\mu_0}{4\pi}\right)^2 \frac{3(2\pi\gamma)^4\hbar^2}{2} s(s+1)[J^{(1)}(\omega_0) + J^{(2)}(2\omega_0)]$$

(Eq. D2.6–2)

and

$$\left[\frac{1}{T_2^{(A)}}\right]_{(rot)} = \left(\frac{\mu_0}{4\pi}\right)^2 \gamma^4\hbar^2\frac{3}{2}s(s+1)$$

$$\times \left[\frac{1}{4}J^{(0)}(0) + \frac{5}{2}J^{(1)}(\omega_0) + \frac{1}{4}J^{(2)}(2\omega_0)\right]$$

(Eq. D2.6–3)

where s denotes the spin of the particles (s = 1/2 of interest here) and $\mu_0$ is the magnetic permeability of vacuum.

In the case of isotropic rotational motions, a model that excellently represents the fluctuations of dipolar interactions in many liquids including water, the spectral density functions are:

$$J^{(0)}(\omega) = \frac{24}{15\beta^6}\left(\frac{\tau_c}{1+\omega^2\tau_{(rot)}^2}\right)$$

(Eq. D2.6–4)

and,

$$J^{(1)}(\omega) = \frac{4}{15\beta^6}\left(\frac{\tau_c}{1+\omega^2\tau_{(rot)}^2}\right)$$

(Eq. D2.6–5)

and,

$$J^{(2)}(\omega) = \frac{16}{15\beta^6}\left(\frac{\tau_c}{1+\omega^2\tau_{(rot)}^2}\right)$$

(Eq. D2.6–6)

where $\tau_{(rot)}$ is the rotational correlation time and $\beta$ is the intramolecular separation between the two spins.

Hence, the rotational $T_1$ relaxation rate is:

$$\left[\frac{1}{T_1^{(A)}}\right]_{(rot)} = \left[\frac{1}{\left(1 + \omega_0^2 \tau_{(rot)}^2\right)} + \frac{4}{\left(1 + 4\omega_0^2 \tau_{(rot)}^2\right)}\right] \lambda_0 \tau_{(rot)}$$

(Eq. D2.6–7)

In a further theoretic development, Torrey (Torrey, 1953) derived an expression for the translational $T_1$ relaxation rate, which is expressed in terms of a translational correlation time:

$$\left[\frac{1}{T_1^{(A)}}\right]_{(trans)} = \left[\frac{1}{\left(1 + \omega_0^2 \tau_{(trans)}^2\right)} + \frac{2}{\left(4 + \omega_0^2 \tau_{(trans)}^2\right)}\right]$$

$$\times PD^{(norm)} \lambda_1 \tau_{(trans)}$$

(Eq. D2.6–8)

This last equation is valid for isotropic diffusion such that the mean molecular flight path is much longer than the inter-molecular proton-to-proton closest distance of approach ($\langle r^2 \rangle \gg a^2$) (see Fig. C.2).

In the equations above (SI units), ($\omega_0$) is the Larmor angular frequency, ($PD^{(norm)}$) is the tissue proton density normalized to pure water at the same temperature; $\rho_w$ the absolute proton density of water expressed as the number of $^1$H-protons per m$^3$, and ($\beta$) is the intra-molecular $^1$H-proton distance. Furthermore, the following quantities have been defined:

$$\lambda_0 \equiv \frac{3}{10} \left(\frac{\mu_0}{4\pi}\right)^2 \frac{(2\pi\gamma)^4 \hbar^2}{\beta^6}$$

(Eq. D2.6–9)

and,

$$\lambda_1 \equiv \frac{2\pi}{5} \left(\frac{\mu_0}{4\pi}\right)^2 \frac{(2\pi\gamma)^4 \hbar^2}{a^3} \rho_w$$

(Eq. D2.6–10)

Altogether, the total kinetic $T_1$ relaxation rate is given by:

$$\left[\frac{1}{T_1^{(A)}}\right]_{(kin)} = \left[\frac{1}{\left(1 + \omega_0^2 \tau_{(rot)}^2\right)} + \frac{4}{\left(1 + 4\omega_0^2 \tau_{(rot)}^2\right)}\right] \lambda_0 \tau_{(rot)}$$

$$+ \left[\frac{1}{\left(1 + \omega_0^2 \tau_{(trans)}^2\right)} + \frac{2}{\left(4 + \omega_0^2 \tau_{(trans)}^2\right)}\right]$$

$$\times PD^{(norm)} \lambda_1 \tau_{(trans)}$$

(Eq. D2.6–11)

The translational correlation time is a function of the tissue viscosity and temperature *via* Stokes equation, specifically

$$\tau_{(\text{trans})} = 4\pi R_{vw}^3 \frac{\eta}{3k_B T} \qquad \text{(Eq. D2.6–12)}$$

where $R_{vw}$ is the van der Waals radius of a molecule in a medium of viscosity $\eta$ at temperature T. Using Stokes-Einstein's formula for viscosity ($\eta = kT/6\pi DR_{vw}$) and Eq. D2.6–11, translational correlation time can be expressed in terms of the diffusion coefficient:

$$\tau_{(\text{tran})} = \left(\frac{2}{9}\right) \frac{R_{vw}^2}{D} \qquad \text{(Eq. D2.6–13)}$$

The extreme narrowing regime refers to situations in which the correlation times are much shorter than the Larmor precession period:

$$\omega_0 \tau_{(\text{rot})} \ll 1 \quad \text{and} \quad \omega_0 \tau_{(\text{tran})} \ll 1 \qquad \text{(Eq. D2.6–14)}$$

in this case, Eq. D2.6–11 reduces to a much simpler form:

$$\left[\frac{1}{T_1^{(A)}}\right]_{(\text{kin})} = 5\lambda_0 \tau_{(\text{rot})} + 3PD^{(\text{norm})}\lambda_1 \tau_{(\text{trans})} \qquad \text{(Eq. D2.6–15)}$$

which is an excellent approximation for pure water.

From a related BPP analysis (Callaghan, 1993), the mathematical steps of which are also beyond the scope of this book, the $T_2$ relaxation rate due to intramolecular interactions is given by:

$$\left[\frac{1}{T_2^{(A)}}\right]_{(\text{rot})} = \frac{1}{2}\left[3 + \frac{5}{(1 + \omega_0^2\tau_{(\text{rot})}^2)} + \frac{2}{(1 + 4\omega_0^2\tau_{(\text{rot})}^2)}\right]\lambda_0 \tau_{(\text{rot})}$$

$$\text{(Eq. D2.6–16)}$$

We note that in the extreme narrowing regime, we have the following equality:

$$\left[\frac{1}{T_1^{(A)}}\right]_{(\text{rot})} = \left[\frac{1}{T_2^{(A)}}\right]_{(\text{rot})} = 5\lambda_0 \tau_{(\text{rot})} \qquad \text{(Eq. D2.6–17)}$$

This partial result would seem to imply that for simple fluids such as water and cerebro-spinal fluid (CSF), $T_1$ is approximately equal to $T_2$. One should be mindful however that even for such simple systems compared to biological tissue, $T_2$ is shorter than $T_1$ and that a more accurate $T_2$ formula should include all kinetic $T_1$ relaxation mechanisms, including the effects of diffusion. Hence, we write:

$$\left[\frac{1}{T_2^{(A)}}\right]_{(kin)} = \frac{3}{2}\lambda_0\tau_{(rot)} + \frac{1}{2}\left[\frac{1}{T_1^{(A)}}\right]_{(kin)} \qquad \text{(Eq. D2.6–18)}$$

where the first term is the secular component of $T_2$:

$$\left[\frac{1}{T_2^{(A,sec)}}\right] = \frac{3}{2}\lambda_0\tau_{(rot)} \qquad \text{(Eq. D2.6–19)}$$

and describes the spectral density that corresponds to the DC component of the randomly fluctuating field, the effect of which is dephasing. The second non-secular term should be calculated by means of Eq. D2.6–11.

In summary, we conclude this section by noticing that to a very good approximation the two relaxation times $T_1$ and $T_2$ of water and simple fluids are frequency independent. These remarkable properties of pure water make it a unique material for qMRI calibration purposes across different scanner platforms of any magnetic field strength in the clinical range. Moreover, the temperature dependence of $T_1$ is known to a great accuracy form experimental data (Hindman, Svirmickas, & Wood, 1973):

$$T_1^{(A)}(T^\circ C) = 10^{-6}T^3 + 7\,10^{-4}T^2 + 0.0595T + 1.6683$$
$$\text{(Eq. D2.6–20)}$$

where $T_1$ is expressed in units of seconds.

The temperature dependence of the relaxation times can also be predicted using simple terms from the temperature dependence of the correlation time, which is modeled by the Arrhenius relation:

$$\tau_{(rot)} = \tau_{(rot)}^0 \exp\left(-\frac{E_a^{(rot)}}{k_B T}\right) \qquad \text{(Eq. D2.6–21)}$$

where $E_a^{(rot)}$ is the activation energy for the rotational molecular motion.

The temperature dependence of the water diffusion coefficient has also been carefully studied experimentally (Simpson & Carr, 1958; Tofts

*(Continued)*

*(Continued)*

---

*et al.*, 2008); to a high degree of accuracy, it can be fitted giving:

$$D(10^{-9} \, m^2 \, s^{-1}) = 10^{-6}D^3 + 4 \, 10^{-4}D^2 + 0.0378D + 1.1145$$

$$\text{(Eq. D2.6–22)}$$

which shows a monotonically increasing behavior with increasing temperature.

---

### D2.7. *Relaxation by Paramagnetic Solutes*

The presence of a paramagnetic solute causes the longitudinal and the transverse relaxation rates of the solvent to increase. Paramagnetic solutes include atoms and ions with unpaired electrons in the outer shell. The relaxation mechanism is dipole-dipole interactions between the $^1$H-protons and the unpaired electrons of the paramagnetic solute. Accordingly, the total relaxations rates are the sum of the diamagnetic relaxation rates and the paramagnetic relaxation rates, *i.e.*

$$\left[\frac{1}{T_{1,2}^{(A)}}\right] = \left[\frac{1}{T_{1,2}^{(A)}}\right]_{(dia)} + \left[\frac{1}{T_{1,2}^{(A)}}\right]_{(par)} \qquad \text{(Eq. D2.7–1)}$$

The diamagnetic relaxation rates include all relaxation processes discussed earlier in this chapter. The paramagnetic contributions to the relaxation rates are proportional to the concentration of paramagnetic compound (Lauffer, 1987); the proportionality constants $r_1$ and $r_2$ are the known as the relaxivities of the paramagnetic material.

$$\left[\frac{1}{T_{1,2}^{(A)}}\right]_{(par)} = r_{1,2}C_{(para)} \qquad \text{(Eq. D2.7–2)}$$

Of interest to qMRI, are the paramagnetic effects of the endogenous dissolved molecular oxygen (Mirhej, 1965; Zaharchuk, Martin, Rosenthal, Manley, & Dillon, 2005) and exogenous gadolinium based contrast agents (Rohrer, Bauer, Mintorovitch, Requardt, & Weinmann, 2005).

### D2.8. *Spin Locking*

In the presence of strong resonantly rotating rf field, a the magnetization of a spin packet will behave as if it were relaxing toward thermal equilibrium in the coordinate frame rotating with such applied field. The reason is that the Hamiltonian including such strong fields $\gamma B_1$ exceeds the natural NMR linewidth is less time dependent in such a reference frame. The corresponding relaxation time, which is known as $T_{1\rho}$ can be calculated also with the perturbation theory/spectral density approach of the BPP theory and such a treatment (Callaghan, 1993) (Look & Lowe, 1966) leads to:

$$\frac{1}{T_{1\rho}} = \left(\frac{\mu_0}{4\pi}\right)^2 \gamma^4 \hbar^2 \frac{3}{2} s(s+1) \left[\frac{1}{4}J^{(0)}(\omega_1) + \frac{5}{2}J^{(1)}(\omega_0) + \frac{1}{4}J^{(2)}(2\omega_0)\right]$$

(Eq. D2.8–1)

where the spectral density functions are given in section D2.6.

## D3. Relaxation in Tissue

### D3.1. *Spatial Scalability and Exponentiality*

As discussed in section D2.6, MR relaxation of bulk water and of paramagnetic solutions are well-understood theoretical problems leading to straightforward equations for the relaxation times in terms of the rotational and translational correlation times. Furthermore, these equations predict frequency independent relaxation times in the extreme narrowing regime, which applies to liquid water in the temperature range of interest. An additional theoretical simplification stems from the fact that when analyzing bulk water and simple fluids, distinguishing between the spatial scales of spin packets relative to the voxel spatial scales is unnecessary. In other words, bulk water and simple fluids are spatially scalable systems.

This is not the case for biological materials and tissue, which are not scalable from the spin packet scale up to the voxel scale: in tissue, water can exist in different geometrical compartments *e.g.* intra- *versus* extra-cellular water as well as in different kinetic states. Two such kinetic states are commonly identified for developing relaxation models in tissue, specifically bound water in the hydration layers of macromolecules and other microscopic structures *e.g.* cell membranes and "free" water in a kinetic state similar to that of bulk water.

Since biological materials are in general not spatially scalable, the question arises as to the interpretation of voxel scale measurements in tissue. In a rudimentary model, the measured relaxation rates at the voxel scale can be interpreted as voxel-averages of the intravoxel distribution of the spin packet relaxation rates, *i.e.*:

$$\frac{1}{T_{1,2(i,j,sl)}} \equiv \left\langle \frac{1}{T_{1,2}(\vec{x})} \right\rangle_{(i,j,sl)} \qquad \text{(Eq. D3.1–1)}$$

where the bracket notation denotes average over the voxel. In such basic scalable theoretical framework, the voxel-scale magnetization dynamics would obey generalized voxel-scale BTS equations of the same mathematical makeup as the spin-packet equations but written in terms of a voxel-average magnetization. This is a common implicit assumption in the literature because the qMRI algorithms used are based on spin-packet scale solutions of the BTS equations.

### D3.2. *Empirical Tissue Relaxometry*

This section follows closely the findings of very comprehensive seminal papers on tissue relaxometry published in the mid eighties (Bottomley, Foster, Argersinger, & Pfeifer, 1984; Cameron, Ord, & Fullerton, 1984; Fullerton, Cameron, & Ord, 1984). When water is associated with macromolecular structures, as in biological tissues, the $T_1$ values are not only much shorter but also dependent on the resonant frequency (Fullerton, *et al.*, 1984). As a rule of thumb, $T_1$ of tissue increases with frequency. The precise dependencies are tissue dependent and in many tissues vary as one-half power of frequency:

$$T_{1(i,j,sl)} \propto \langle \nu(\vec{x}) \rangle_{(i,j,sl)}^{1/2} \qquad \text{(Eq. D3.2–1)}$$

One of the above mentioned articles, summarizes their own findings as well as the information in the literature available at the time; the conclusions based on the study of 15 rat tissues are reproduced with only minor textual modifications in the following (Cameron, *et al.*, 1984):

1. Each normal tissue type is characterized by reproducible values of the $T_1$ and $T_2$ relaxation time.
2. In comparison to pure water or dilute electrolyte solutions, the proton $T_1$ relaxation times of biological tissue are much reduced; these authors (Cameron, *et al.*, 1984) give the following range of reduction factors: 2.5 to 17 times. In addition, the $T_2$ relaxation times are even more reduced relative to pure water: in this case these authors give a $T_2$ reduction range of 19 to more than 80 times. In each tissue, the $T_1$ was consistently longer than the $T_2$.

3. A significant positive correlation exists between tissue $T_1$ relaxation times and tissue water content. In other words, this relationship is such that tissues with higher water content tend to have longer $T_1$ times. In contrast, tissue $T_2$ times are not as well correlated to water content, as are tissue $T_1$ times.

4. Tissues with shorter $T_1$ have higher fractions of hydration water, larger amounts of rough endoplasmic reticulum, and greater protein synthetic rates.

5. Tissues with more non-membrane bounded lipid droplets tend to have longer $T_2$ relaxation times. This accounts for the the longer $T_2$ of fat and adipose relative to aqueous tissues.

6. In general, the shorter the $T_2$, the greater the surface area encountered by diffusing tissue water. This is whether the surface area is the form of cell membranes or in the form of intracellular or extracellular filamentous or fibrillar molecules. In short, $T_2$ appears to be inversely related to total surface area encountered by water molecules.

7. Tissues with higher water content tend to have longer $T_1$ and $T_2$ values, large amounts of fluid-filled luminal space, and/or are deficient in overall tissue surface areas.

Deviations from the general rules above (Cameron, *et al.*, 1984) offer insights into understanding the increased $T_1$ and $T_2$ values which typically occur when a healthy tissue (*e.g.* liver parenchyma) undergoes malignant transformation into a tumor (*e.g.* hepatoma).

## References

Abragam, A. (1961). *The Principles of Nuclear Magnetism*: Clarendon Press Oxford.

Bloembergen, N. (1957). Proton relaxation times in paramagnetic solutions. *The Journal of Chemical Physics, 27*, 572.

Bloembergen, N., Purcell, E. M., & Pound, R. V. (1948). Relaxation effects in nuclear magnetic resonance absorption. *Physical Review, 73*(7), 679–712.

Bottomley, P., Foster, T., Argersinger, R., & Pfeifer, L. (1984). A review of normal tissue hydrogen NMR relaxation times and relaxation mechanisms from 1–100 MHz: dependence on tissue type, NMR frequency, temperature, species, excision, and age. *Medical Physics, 11*, 425.

Callaghan, P. (1993). *Principles of Nuclear Magnetic Resonance Microscopy*: Oxford University Press, USA.

Cameron, I., Ord, V., & Fullerton, G. (1984). Characterization of proton NMR relaxation times in normal and pathological tissues by correlation with other tissue parameters* 1. *Magnetic resonance imaging, 2*(2), 97–106.

Carr, H., & Purcell, E. (1954). Effects of diffusion on free precession in nuclear magnetic resonance experiments. *Physical Review, 94*(3), 630–638.

Dixon, R. L., & Ekstrand, K. E. (1982). The physics of proton NMR. *Med Phys, 9*(6), 807–818.

Fullerton, G., Cameron, I., & Ord, V. (1984). Frequency dependence of magnetic resonance spin-lattice relaxation of protons in biological materials. *Radiology, 151*(1), 135.

Hindman, J., Svirmickas, A., & Wood, M. (1973). Relaxation processes in water. A study of the proton spin lattice relaxation time. *The Journal of Chemical Physics, 59*, 1517.

Kubo, R., & Tomita, K. (1954). A general theory of magnetic resonance absorption. *Journal of the Physical Society of Japan, 9*, 888.

Lauffer, R. (1987). Paramagnetic metal complexes as water proton relaxation agents for NMR imaging: theory and design. *Chem Rev, 87*(5), 901–927.

Look, D., & Lowe, I. (1966). Nuclear Magnetic Dipole—Dipole relaxation along the static and rotating magnetic fields: application to gypsum. *The Journal of Chemical Physics, 44*, 2995.

Mirhej, M. (1965). Proton spin relaxation by paramagnetic molecular oxygen. *Canadian Journal of Chemistry, 43*(5), 1130–1138.

Rohrer, M., Bauer, H., Mintorovitch, J., Requardt, M., & Weinmann, H. (2005). Comparison of magnetic properties of MRI contrast media solutions at different magnetic field strengths. *Investigative Radiology, 40*(11), 715.

Simpson, J., & Carr, H. (1958). Diffusion and nuclear spin relaxation in water. *Physical Review, 111*(5), 1201–1202.

Slichter, C. (1990). *Principles of Magnetic Resonance*: Springer Verlag.

Solomon, I. (1955). Relaxation processes in a system of two spins. *Physical Review, 99*(2), 559–565.

Thomas, S. (1986). NMR in medicine: the instrumentation and clinical applications. *AAPM Medical Physics Monograph No., 14*.

Tofts, P. S., Jackson, J. S., Tozer, D. J., Cercignani, M., Keir, G., MacManus, D. G., *et al.* (2008). Imaging cadavers: cold FLAIR and noninvasive brain thermometry using CSF diffusion. *Magn Reson Med, 59*(1), 190–195.

Torrey, H. C. (1953). Nuclear spin relaxation by translational diffusion. *Physical Review, 92*(4), 962–969.

Zaharchuk, G., Martin, A., Rosenthal, G., Manley, G., & Dillon, W. (2005). Measurement of cerebrospinal fluid oxygen partial pressure in humans using MRI. *Magnetic Resonance in Medicine, 54*(1), 113–121.

# E. QMRI THEORY

## E1. Introduction

This chapter is concerned the general principles of qMRI and the specific theoretico-practical paradigms for mapping individual qMRI parameters following the order of section C5.6.

## E2. qMRI Principles

A qMRI technique consists of two complementary components that should be designed as a matched pair, specifically a qMRI pulse sequence and a qMRI algorithm. The former is used to generate the directly acquired (DA) images that are post-processed with the qMRI algorithm for generating the qMRI maps.

qMRI pulse sequences are variants of standard MRI pulse sequences, which are modified and optimized for the purpose of generating images that are suitable for qMRI processing. A very fruitful line of qMRI techniques stem from the principle of differential weighting whereby a qMRI parameter is mapped from images that were acquired under identical experimental conditions, except for the degree of weighting to the qMRI parameter to be mapped. In other words, the only difference between two differentially weighted images is the value of the qCV that controls the degree of weighting the targeted qMRI parameter. Accordingly, the most useful qMRI pulse sequences can efficiently generate high quality "differentially-weighted datasets" in clinically compatible scan times.

This chapter is concerned with the theoretical basis and the practical considerations pertaining to the design of qMRI techniques, from image acquisition to qMRI map generation. We include a streamlined review of standard MRI pulse sequences, with focus on those that are particularly suited for qMRI.

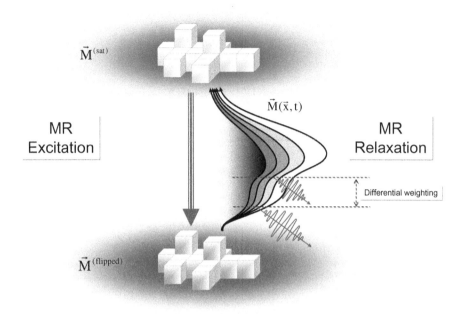

Fig. E-1. Diagram illustrating a one TR cycle of an MRI pulse sequence. The magnetization vector is forced to cycle between the saturated value $\vec{M}^{\text{sat}}$ and $\vec{M}^{\text{flipped}}$. During the excitatory part of the TR cycle, tissues are removed from the state of dynamic equilibrium. During the relaxation portion of the TR cycle, tissue magnetizations return towards equilibrium along trajectories that depend on the targeted qMRI parameter. By acquiring two (or more) data sets with different weightings, a map of the targeted qMRI can be calculated.

The principles of qMRI stem from the pixel value equation and the BTS equations. Three principles can be readily formulated: 1) the principle of weighting reversal with pixel value calibration for quantifying the proton density of the liquid pool, 2) the principle of differential weighting for quantifying the liquid pool qMRI parameters, and 3) the principle of differential weighting for the semisolid pool parameters. We draw a difference between the principles of differential weighting for the liquid pool *versus* the semisolid pool because the qMRI parameters of the liquid pool can be determined individually and the parameters of the semisolid pool are calculated simultaneously as a group from a set of magnetization transfer weighted images.

## E2.1. *The Pixel Value Equation (Revisited)*

From a mathematical perspective, qMRI parameters are continuous variables that occur intertwined with pulse sequence qCVs in the MRI pixel value equation. As stated above, most qMRI techniques consist of several MRI acquisitions obtained with different values of a specific qCV. We will use the notation: qCV_X($\lambda$) to designate a qCV named "X" with specific value "lambda"; for example, for a multi echo pulse sequence that acquires n spin-echoes, X = TE and the qCV values are:

$$\text{qCV\_X}(\alpha) = \text{qCV\_TE}(1), \text{qCV\_TE}(2), \ldots \text{qCV\_TE}(n) \quad \text{(Eq. E2.1--1)}$$

Using this notation, we rewrite the pixel value equation of acquisition in the compact form:

$$\text{pv}_{(i,j,sl)}^{\text{Acq}\_\lambda} = \Gamma_{(i,j,sl)} \text{PD}_{(i,j,sl)}^{(A)} \text{PSw}(\cdots \text{qCV\_X}(\lambda) \cdots || \cdots \text{qMRI\_par}_{(i,j,sl)} \cdots) \quad \text{(Eq. E2.1--2)}$$

where the full pulse sequence weighting factor includes the $T_2$ decay factor, *i.e.*

$$\text{PSw}(\cdots \text{qCV\_X}(\lambda) \cdots || \cdots \text{qMRI\_par}_{(i,j,sl)} \cdots)$$

$$= \text{psw}(\cdots \text{qCV\_X}(\lambda) \cdots || \cdots \text{qMRI\_par}_{(i,j,sl)} \cdots) \exp\left(-\frac{\text{TE}}{T_{2,(i,j,sl)}^{(A)*}}\right) \quad \text{(Eq. E2.1--3)}$$

The reader is reminded that $T_2^{(A)}$ should be used instead of $T_2^{(A)*}$ in Eq. E2.1-3 for spin-echo sequences.

In addition, in Eq. E2.1-2, we have combined all non-dynamic factors into in one factor given by

$$\Gamma_{(i,j,sl)} \equiv \left(\frac{\pi^2 \gamma^2 \hbar^2 B_0}{k_B T}\right) \Omega_{(i,j,sl)} \Delta V_{(i,j,sl)} \quad \text{(Eq. E2.1--4)}$$

This factor encompasses the experimental conditions of field strength, temperature, receiving sensitivity, and geometry *via* the voxel volume.

Additionally, qMRI parameters can occur in the magnitude or in the phase of the pulse sequence weighting factor, which as shown in Section C5.5

is in general complex-valued:

$$\text{PSw}(\cdots \text{qCV\_X}(\lambda) \cdots \| \text{qMRI\_par}_{(i,j,sl)})$$

$$= |\text{PSw}(\cdots \text{qCV\_X}(\lambda) \cdots \| \cdots \text{qMRI\_par}_{(i,j,sl)})|$$

$$\times \exp(i\Phi_w^{(\text{App})}(\cdots \text{qCV\_X}(\lambda) \cdots \| \text{qMRI\_par}_{(i,j,sl)})) \quad (\text{Eq. E2.1–5})$$

thus leading to two distinct types of qMRI techniques, specifically magnitude and phase techniques. The solutions of the BTS equations dictate that magnitude-qMRI-techniques lend themselves naturally to qMRI of the relaxation times, the diffusion tensor, perfusion, as well as the MT parameters that are associated to the semisolid pool. The BTS solutions also imply that phase-qMRI-techniques lend themselves naturally to quantifying the velocity of flowing and moving spin packets.

### E2.2. *Association of qMRI Parameters with Weighting qCVs: Liquid Pool*

As discussed in section C5.6, the qMRI parameters of the liquid pool can be grouped into three parameter categories, specifically: 1) the proton density, which is unique and representative of $^1$H-proton content. 2) The kinetic parameters of tissue diffusion and perfusion, and 3) the relaxation time parameters that are indicative of the magnetic interactions between $^1$H-protons with each other and with the microenvironment. Altogether and in tabular form, the qMRI parameters for the directly detectable or MR-visible liquid pool are:

$$\text{qMRI\_par}^{(A)} = \begin{Bmatrix} \text{PD}^{(A)} & \cdots & \cdots & \cdots \\ \vdots & \mathbf{D}^{(A)} & (f^{(A)}, D^{*(A)}) & \vec{v}^{(A)} \\ \vdots & T_1^{(A)} & (T_2^{(A)}, T_2^{*(A)}) & T_{1,2\rho}^{(A)} \end{Bmatrix} \quad (\text{Eq. E2.2–1})$$

In general, qMRI pulse sequences are designed such that the qMRI parameter of the liquid pool that is targeted for mapping occurs in expressions as multiplication or division factors of a pulse sequence qCV, which is the weighting qCV. Accordingly, the liquid pool qMRI parameter and the pulse sequence qCV can be thought of as a pair of qMRI associated variables: for example, the qMRI pair formed by $T_2$ and TE (spin-echo).

Other qMRI-associated variables are as follows:

$$
\text{qMRI\_par}^{(A)} = \left\{ \begin{matrix} \text{PD}^{(A)} & \cdots & \cdots & \cdots \\ \vdots & \mathbf{D}^{(A)} & (f^{(A)}, D^{*(A)}) & \vec{v}^{(A)} \\ \vdots & T_1^{(A)} & (T_2^{(A)}, T_2^{*(A)}) & T_{1,2\rho}^{(A)} \end{matrix} \right\}
$$

$$\Uparrow$$

qMRI Association $\qquad\qquad$ (Eq. E2.2–2)

$$\Downarrow$$

$$
\text{qCV\_X} = \left\{ \begin{matrix} \text{na} & \cdots & \cdots & \cdots \\ \vdots & \text{b-matrix} & b_{x,y,z} & \vec{V}^{(\text{enc})} \\ \vdots & (\text{TR}, \text{TI}, \text{FA}) & (\text{TE}^{(\text{SE})}, \text{TE}^{(\text{GE})}) & \text{TSL} \end{matrix} \right\}
$$

As will be explained on a case-by-case basis in the next section, the qCVs that are associated with the kinetic qMRI parameters stem from using pairs of balanced gradient pulses in different configurations; such pulses can be used either with gradient-echo or with spin-echo pulse sequences. Gradient-echo pulse sequences are often used for quantifying flow and spin-echo --or spin-echo like-- pulse sequences are used for diffusion qMRI. Also, the qCV for quantifying the relaxation times in the rotating frame of reference $(T_{1,2,\rho}^{(A)})$ is the duration of the spin locking rf pulse.

We note again that the proton density is not associated with any known weighting qCV and therefore this unique and fundamental qMRI parameter may not be quantified *via* differential weighting, as explained in the following sections of this chapter. Furthermore, some qMRI parameters have more than one associated qCVs as is the case of $T_1$, which depending on the specific qMRI pulse sequence, may be quantified by differential weighting *via* the repetition time TR, the inversion time TI, or the excitation flip angle FA.

## E2.3. *The Principle of Differential Weighting*

The principle of differential weighting applies to several images that are acquired under identical experimental conditions with the exception of the value of a particular qCV, which is associated to the qMRI parameter targeted for quantification (see section E2.2). This principle states that differences in weightings between two (or more) DA images can be used to compute the qMRI parameter that is responsible for the observed differential weighting on a pixel-by-pixel basis. Furthermore, the mathematical algorithms used in these computations are greatly simplified by exploiting the multiplicative structure of pixel value equation above, which implies that the pixel ratios between differentially weighted images are independent of the experimental conditions and depend only on the ration of the pulse sequence weighting factors.

$$\frac{pv_{(i,j,sl)}^{Acq\_1}}{pv_{(i,j,sl)}^{Acq\_2}} = \frac{PSw(\cdots qCV\_X(1)\cdots || \cdots qMRI\_par_{(i,j,sl)}\cdots)}{PSw(\cdots qCV\_X(2)\cdots || \cdots qMRI\_par_{(i,j,sl)}\cdots)} \qquad \text{(Eq. E2.3–1)}$$

Hence the practical utility of the principle of differential weighting: by modeling the pulse sequence weighting factors with MR physics, the desired tissue property can be mapped *via*:

$$qMRI\_par_{(i,j,sl)} = \Re\left(\frac{pv_{(i,j,sl)}^{Acq\_1}}{pv_{(i,j,sl)}^{Acq\_2}}, qCV\_X(1), qCV\_X(2)\right) \qquad \text{(Eq. E2.3–2)}$$

where the symbol $\Re$ is a function or algorithm derived by solving the BTS equations that describe the pulse sequence used to generate the DA images.

As a concrete example, we illustrate the formulas above by analyzing the simple case of dual-echo spin-echo acquisitions; in this case, the pixel value ratio is given by the simple expression

$$\frac{pv_{(i,j,sl)}^{Acq\_1}}{pv_{(i,j,sl)}^{Acq\_2}} = \frac{\exp\left(-\frac{TE1}{T_{2,(i,j,sl)}}\right)}{\exp\left(-\frac{TE2}{T_{2,(i,j,sl)}}\right)} \qquad \text{(Eq. E2.3–3)}$$

which leads to an analytical solution for mapping $T_2$, specifically:

$$T_{2,(i,j,sl)} = \frac{(TE2 - TE1)}{\ln\left(\frac{pv_{(i,j,sl)}^{Acq\_1}}{pv_{(i,j,sl)}^{Acq\_2}}\right)} \qquad \text{(Eq. E2.3–4)}$$

It is instructive to write this solution in a more general form, which will be useful when dealing with cases that are more complicated because analytical solutions may not exist. To this end, we note that Eq 2.3-4 is equivalent to the null equation:

$$f(T_{2,(i,j,sl)}) = T_{2,(i,j,sl)} - \frac{(TE2 - TE1)}{\ln\left(\frac{pv_{(i,j,sl)}^{Acq\_1}}{pv_{(i,j,sl)}^{Acq\_2}}\right)} = 0 \qquad \text{(Eq. E2.3–5)}$$

and therefore, $T_2$ can be found as a root of the function f on the left hand side of Eq. E2.2-5, specifically:

$$T_{2,(i,j,sl)} = \text{Root} \left( T_{2,(i,j,sl)} - \frac{(TE2 - TE1)}{\ln \left( \frac{pv_{(i,j,sl)}^{Acq\_1}}{pv_{(i,j,sl)}^{Acq\_2}} \right)}, \tau_2 \right) \qquad \text{(Eq. E2.3–6)}$$

In this expression, $\tau_2$ is an initial guess value of the solution. This is a trivial example of the problem embodied by Eq. E3.4-2, which has general validity even when analytical solutions are not available. Moreover, Eq. E3.4-6 establishes a connection between qMRI and root finding problems, the theory of which is a well-established branch of mathematics; root finding algorithms are standard built-in functions of most high level programming languages and environment. These are typically built around the secant/Mueller and the Rider and Brent root finding methods and many others exist (Press, Teukolsky, Vetterling, & Flannery, 1993).

## E3. Parameter Specific qMRI Paradigms

### E3.1. *qMRI of the Proton Density: Un-weighting and Pixel Value Calibration*

Since the proton density is not paired to any known pulse sequence qCV, it is not currently amenable to quantification by differential weighting and therefore alternative means of quantification are needed. Solving for the proton density, the pixel value equation reads:

$$PD_{(i,j,sl)}^{(A)} = \Gamma_{(i,j,sl)}^{-1} \frac{pv_{(i,j,sl)}^{Acq\_\lambda}}{PSw(\cdots qCV\_X(\lambda) \cdots || \cdots qMRI\_par_{(i,j,sl)} \cdots)}$$

$$\text{(Eq. E3.1–1)}$$

Accordingly, in principle the proton density can be calculated on a pixel-wise basis by dividing the measured pixel values by the product of the experimental conditions factor $\Gamma$ and the pulse sequence weighting factor of each voxel.

In turn, the pulse sequence weighting factor can be calculated if all the qMRI parameters –typically $T_1$, $T_2$, and D-- that weight the specific pixel value are known at each pixel location. In general, this process of weighting reversal or un-weighting the pixel values is possible with multispectral qMRI techniques, with which self-coregistered maps of the main qMRI parameters can be generated from a single MRI acquisition, as will be discussed in detail later.

Determining the experimental conditions factor is however a more delicate undertaking mainly because of the receiving sensitivity factor, which is position dependent and not readily available. The voxel volume of Fourier transform MRI on the other hand is position-independent to a very good approximation, if the gradient subsystem does not exhibit significant deviations from linearity, hence

$$\Gamma^{-1}_{(i,j,sl)} = \left( \frac{k_B T}{\pi^2 \gamma^2 \hbar^2 B_0 \Delta V} \right) \Omega^{-1}_{(i,j,sl)} \qquad \text{(Eq. E3.1–2)}$$

and the main problem is determining the receiving sensitivity distribution in three dimensions. For most quadrature coils as well as for modern phased-array coil systems, the spatial variations of the receiving system are relatively small and can be determined in relative terms by scanning separately a homogeneous phantom. Furthermore, an amplitude normalization point is needed, which is conveniently chose as a pixel containing a pure fluid, such as cerebrospinal fluid, which has a normalized $PD^{(A)}_{(cal)} = 1$. With these considerations, one can normalize to the value of maximum proton density, i.e. that of pure water,

$$\Omega_{(i,j,sl)} = \Omega^{water}_{(cal)} S_{(i,j,sl)} \qquad \text{(Eq. E3.1–3)}$$

where $S_{(i,j,sl)}$ is the dimensionless sensitivity function that has been normalized to water.

$$PD^{(A)}_{(i,j,sl)} = \left( \frac{k_B T}{\pi^2 \gamma^2 \hbar^2 B_0 \Delta V} \right)$$

$$\times \frac{pv^{Acq\_\lambda}_{(i,j,sl)}}{\Omega^{water}_{(cal)} S_{(i,j,sl)} PSw(\cdots qCV\_X(\lambda) \cdots || \cdots qMRI\_par_{(i,j,sl)} \cdots)}$$

$$\text{(Eq. E3.1–4)}$$

As written above, the pixel value as well as the pulse sequence weighting factor are complex numbers and the ratio is a real number. In practice it is easier to work the modulus, *i.e.*,

$$PD^{(A)}_{(i,j,sl)} = \left( \frac{k_B T}{\pi^2 \gamma^2 \hbar^2 B_0 \Delta V} \right)$$

$$\times \left| \frac{pv^{Acq\_\lambda}_{(i,j,sl)}}{\Omega^{water}_{(cal)} S_{(i,j,sl)} PSw(\cdots qCV\_X(\lambda) \cdots || \cdots qMRI\_par_{(i,j,sl)} \cdots)} \right|$$

$$\text{(Eq. E3.1–5)}$$

With the procedures described above, all factors and parameters on the right hand side of this equation are available, and therefore this formula can be used for mapping of the absolute proton density where each pixel value expressed in units of number of $^1$H-protons per m$^3$. Many researchers (P Tofts, 2003) prefer to normalize the proton densities to that of pure water at the same temperature, *i.e.*

$$rPD^{(A)}_{(i,j,sl)} = \frac{PD^{(A)}_{(i,j,sl)}}{PD^{(w)}} \qquad \text{(Eq. E3.1–6)}$$

In practice, this can be accomplished either by identifying pixels corresponding to a pure fluid in the image approximating water, ideally pixels of voxels containing CSF, or by using a vial of water next to the patient during the scan. Such relative proton densities are reported in percent units (pu).

The proton density of water, expressed in molar, is in turn related to the mass density of water by

$$PD^{(w)} = \frac{NP}{MW}\rho_w(T) \qquad \text{(Eq. E3.1–7)}$$

where NP is the number of $^1$H-protons per molecule –*i.e.* 2 for water — and MW is the molecular weight –18.0152 amu for water--. The mass density of pure water in kg m$^{-3}$ as a function of temperature in °C is well known and given by the experimental formula (see Fig. E-2):

$$\rho_w(T) \cong 10^3(3.651\,10^{-8}T^3 - 7.439\,10^{-6}T^2 + 4.862\,10^{-5}T + 0.99998)$$
$$\text{(Eq. E3.1–8)}$$

Using the two equations above, the $^1$H-proton densities water at room and body temperature are:

$$PD^{(w)} = \begin{cases} 110.7\,\text{M} & \text{at T} = 24°\text{C} \\ 110.3\,\text{M} & \text{at T} = 37°\text{C} \end{cases} \qquad \text{(Eq. E3.1–9)}$$

and temperature differences should be considered when comparing *in vivo* data to phantom data, which is normally acquired at room temperature (20–25°C).

As shown in the example below, a proton density map is visually very similar to a PD-weighted image obtained with very short TE and very long TR; in this particular example, CSF is brighter in the PD map than in the PD-weighted image, which has some residual CSF signal saturation.

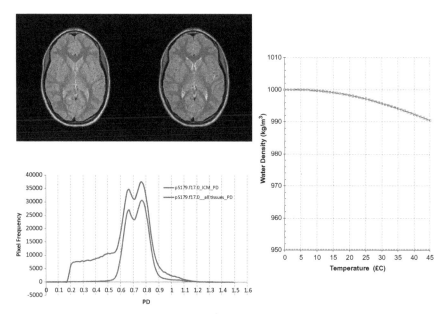

Fig. E-2.   Proton density qMRI. A proton density weighted brain image (top left) is very similar to a PD map (top right), except in this case that the intra-ventricular CSF is noticeably darker in the PD-weighted image because of residual $T_1$-weighting. The whole-head PD histogram is compared to the brain PD histogram (bottom left graph). The effect of temperature on water density is shown in the graph of the right hand side.

### E3.2.   *qMRI of Diffusion and Perfusion: Intravoxel Incoherent Motion (IVIM)*

Magnetic resonance-derived diffusion coefficients are almost exclusively generated by diffusion-weighting (DW) using pulsed-field gradient (PFG) pulse sequences that include balanced magnetic field gradient pulses. This leads to the concept of PFG diffusion weighting whereby the corresponding pulse sequence qCV is known as the b-factor. For a spin-echo pulse sequence equipped with rectangular diffusion weighting pulses, the b-values along any given direction can be calculated with the Stejskal-Tanner formula (Stejskal & Tanner, 1965):

$$\mathbf{b}^{Acq-\lambda} = (2\pi\gamma)^2 g_\lambda^2 \delta^2 \left[ \Delta - \frac{\delta}{3} \right] \qquad \text{(Eq. E3.2–1)}$$

The parameters $g_\lambda, \delta, \Delta$ are the maximum gradient amplitude, the pulse duration, and the inter pulse delay respectively.

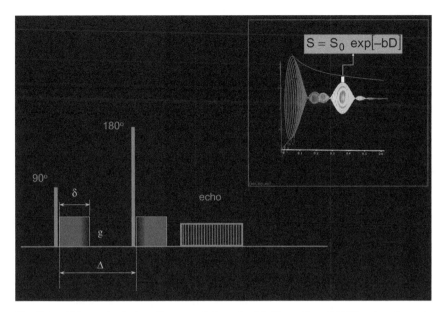

Fig. E-3. Schematic timing diagram of the pulsed field gradient (PFG) spin-echo pulse sequence. The diffusion sensitization pulses define the diffusion qCV, namely the b-factor, which determines the level of diffusional spin-echo attenuation caused by the application of the diffusion gradients (top right insert).

The method of PFG diffusion encoding with balanced gradient pulses was originally conceived for diffusion NMR of whole samples (Stejskal & Tanner, 1965). It was later applied to mapping with diffusion weighted imaging (DWI) the diffusion coefficients along one or several directions (Le Bihan & Breton, 1985; Merboldt, Hanicke, & Frahm, 1985; Taylor & Bushell, 1985), and more recently for mapping all six independent components of the diffusion tensor (P. J. Basser, Mattiello, & LeBihan, 1994; Mattiello, Basser, & Le Bihan, 1997; Pierpaoli, Jezzard, Basser, Barnett, & Di Chiro, 1996). Diffusion tensor imaging (DTI) will be studied in the next section; in this section, we focus on organs and tissues exhibiting isotropic diffusion as well as isotropic perfusion. More specifically, we describe DWI — also known as intravoxel incoherent motion (IVIM) imaging--, which is a method initially developed by Le Bihan *et al.* (Le Bihan *et al.*, 1988) to quantitatively assess the microscopic translational motions that occur in each image voxel and that includes pure molecular water diffusion as well as microcirculation in the capillary bed, or blood perfusion. For such isotropic case, we will ignore the directional aspects of diffusion; accordingly, the pixel value equation takes the form:

$$pv_{(i,j,sl)}^{Acq\text{-}\lambda}$$

$$= \Gamma_{(i,j,sl)}PD_{(i,j,sl)}^{(A)}\ PSw\left(\cdots b^{Acq\text{-}\lambda}\cdots||\cdots f_{(i,j,sl)}^{(A)}, D_{(i,j,sl)}^{(A)}, D_{(i,j,sl)}^{*(A)}\cdots\right)$$

$$(\text{Eq. E3.2-2})$$

where, the pulse sequence weighting factor for a voxel at position (i, j, k) is given by,

$$PSw\left(\cdots b^{Acq\text{-}\lambda}\cdots||\cdots f_{(i,j,sl)}^{(A)}, D_{(i,j,sl)}^{(A)}, D_{(i,j,sl)}^{*(A)}\cdots\right)$$

$$= \exp\left[-\frac{t}{T_2^{(A)}}\right]\left[\left(1 - f_{(i,j,sl)}^{(A)}\right)\exp\left(-b^{Acq\text{-}\lambda}D_{(i,j,sl)}^{(A)}\right) + f_{(i,j,sl)}^{(A)}\right.$$

$$\left.\times \exp\left(-b^{Acq\text{-}\lambda}(D_{(i,j,sl)}^{(A)} + D_{(i,j,sl)}^{*(A)})\right)\right] \qquad (\text{Eq. E3.2-3})$$

This formula describes the signal attenuation due to diffusion and perfusion in diffusion gradients applied in any direction, for a tissue represented by the following three qMRI parameters:

(1) $f_{(i,j,sl)}^{(A)}$ denotes the fraction of $^{1}$H-protons in the capillary bed or perfusion fraction,

(2) $D_{(i,j,sl)}^{(A)}$ is the diffusion coefficient of water molecules in the extra vascular space --also known as the slow component of diffusion--, and

(3) $D_{(i,j,sl)}^{*(A)}$ is the diffusion parameter representing microcirculation, which is incoherent because the capillary bed consists of vessels with pseudorandom orientations; for this reason $D_{(i,j,sl)}^{*(A)}$ is also referred to as the pseudodiffusion coefficient. Furthermore, because blood flow in the capillaries is much faster than water diffusion, this last qMRI parameter is also known as the fast component of diffusion. Luciani et al. (Luciani et al., 2008) used this technique for abdominal imaging and found that for normal adult liver, the diffusion coefficient of the fast perfusion component is about 70–85 times larger than that of the slow component that corresponds to pure molecular diffusion. Moreover, the reported liver perfusion fraction is about 0.26. As noted by Le Bihan (Le Bihan, 2008): "Although the difference in spatial scale between the processes of diffusion (nanometers) and pseudodiffusion (tens of micrometers) extends across five orders of magnitude, it is amazing to observe that the associated diffusion and pseudodiffusion coefficients differ only by roughly one order of magnitude (D, the molecular diffusion coefficient of water in tissues, is about $1 \times 10^{-3}\,\text{mm}^2/\text{sec}$,

while D*, the pseudodiffusion coefficient associated with blood flow, is about $10 \times 10^{-3}\,\text{mm}^2/\text{sec}$ in the brain and $70 \times 10^{-3}\,\text{mm}^2/\text{sec}$ in the liver. This is because those coefficients combine effects of elementary particle velocity and distance. Molecular diffusion is a very fast process, as far as molecular distances are concerned, while blood flow pseudodiffusion is comparatively much slower but is over distances of tens of micrometers."

### E3.2.i. *Diffusion Tensor Imaging (DTI)*

qMRI of the diffusion tensor stems for the generalized pixel value equation that describes tissues with anisotropic diffusion, specifically:

$$
\text{pv}_{(i,j,sl)}^{\text{Acq}\_\lambda} = \Gamma_{(i,j,sl)}\text{PD}_{(i,j,sl)}^{(A)} \exp\left[ -\frac{t}{T_{2(i,j,sl)}^{(A)}} \right]
$$

$$
\times \exp\left[ -\sum_{l,m=x,y,z} (D_{l,m}^{(\text{eff}\_A)})_{(i,j,sl)} b_{l,m}^{\text{Acq}\_\lambda} \right] \quad \text{(Eq. E3.2–4)}
$$

Accordingly, the strategy is to acquire a set of differentially weighted images by applying PFG gradients along several directions such that the six independent components of the diffusion tensor can be calculated on a pixel by pixel basis from a the generated pixel values and the known elements b-matrix of the pulse sequence. Calculating the b-matrix as a function of pulse sequence parameters can lead to complicated equations due to the occurrence of the so-called "cross terms" between the diffusion and the imaging gradient pulses. Minimizing the cross terms can be accomplished by refocusing all imaging gradients just after these have been applied (P. Basser & Pierpaoli, 1998); for such pulse sequences, the pixel value equation simplifies to:

$$
\text{pv}_{(i,j,sl)}^{\text{Acq}\_\lambda} = \Gamma_{(i,j,sl)}\text{PD}_{(i,j,sl)}^{(A)} \exp\left[ -\frac{t}{T_{2(i,j,sl)}^{(A)}} \right] \exp[-\alpha^2 g_0^2 \hat{r}^{\lambda^{\text{T}}} \cdot \mathbf{D}_{(i,j,sl)} \cdot \hat{r}^{\lambda}]
$$

$$
\text{(Eq. E3.2–5)}
$$

implying that diffusion signal attenuation in anisotropic tissue depends on the projection of the diffusion tensor along the diffusion sensitization direction $(\hat{r}^{\lambda^{\text{T}}} \cdot \mathbf{D}_{(i,j,sl)} \cdot \hat{r}^{\lambda})$ used in the specific acquisition. Moreover, in this equation, $g_0$ is the amplitude of the diffusion sensitization pulses and

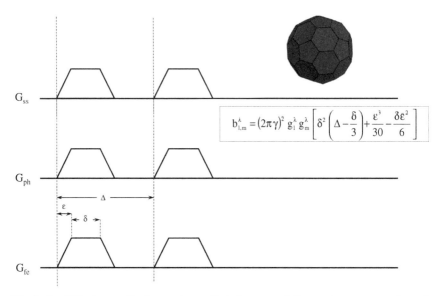

Fig. E-4.   Diffusion gradients may be applied along any of three pulse sequence channels thus providing diffusion sensitization along the physical directions (ss, ph, and fe) of the pulse sequence. Furthermore, by applying gradient pulse along two or three channels simultaneously, diffusion sensitization along any direction can be achieved (insert equation). This allows for generating diffusion-weighted data that is sufficient for quantifying the full diffusion tensor, *i.e.* for quantifying the six independent components of the diffusion tensor. Infinite arrangements are possible: for illustrative purposes, the example of a truncated icosahedron is shown in the top right corner.

$\hat{r}^\lambda$ is a unitary vector pointing in the directions of the gradient field of the specific acquisition ($\lambda$). Furthermore, is a constant that depends on the gyromagnetic ratio, as well as the shapes and the timing parameters of the diffusion sensitizing pulses; for trapezoidal pulses (see Fig. E-4):

$$b_{l,m}^\lambda = (2\pi\gamma)^2 g_l^\lambda\, g_m^\lambda \left[ \delta^2 \left( \Delta - \frac{\delta}{3} \right) + \frac{\varepsilon^3}{30} - \frac{\delta\varepsilon^2}{6} \right] \qquad \text{(Eq. E3.2–6)}$$

and therefore,

$$\alpha^2 \equiv (2\pi\gamma)^2 \left[ \delta^2 \left( \Delta - \frac{\delta}{3} \right) + \frac{\varepsilon^3}{30} - \frac{\delta\varepsilon^2}{6} \right] \qquad \text{(Eq. E3.2–7)}$$

Images suitable for DTI analysis are usually acquired using two gradient amplitudes $(0, g_0)$ and each set of DW images is acquired with a different gradient direction;

$$\vec{g}^{(\text{Acq}\_\lambda)} = g_0 \hat{r}^{(\lambda)} \quad \text{where} \quad \lambda \in \{0, \dots, n\} \qquad \text{(Eq. E3.2–8)}$$

As a concrete example, let us consider the following six diffusion sensitization directions:

$$\hat{r}^{(1)} = \left(\frac{1}{\sqrt{2}}, \ 0, \ \frac{1}{\sqrt{2}}\right)^{T}; \ \ \hat{r}^{(2)} = \left(\frac{-1}{\sqrt{2}}, \ 0, \ \frac{1}{\sqrt{2}}\right)^{T}; \ \ \hat{r}^{(3)} = \left(0, \ \frac{1}{\sqrt{2}}, \ \frac{1}{\sqrt{2}}\right)^{T} \Big\}$$

$$\hat{r}^{(4)} = \left(0, \ \frac{1}{\sqrt{2}}, \ \frac{-1}{\sqrt{2}}\right)^{T}; \ \ \hat{r}^{(5)} = \left(\frac{1}{\sqrt{2}}, \ \frac{1}{\sqrt{2}}, \ 0\right)^{T}; \ \ \hat{r}^{(3)} = \left(\frac{-1}{\sqrt{2}}, \ \frac{1}{\sqrt{2}}, \ 0\right)^{T} \Big\}$$

(Eq. E3.2–9)

as has been used (P. Basser & Pierpaoli, 1998) in a seven-acquisition DTI protocol, which also includes one acquisition without diffusion sensitization.

$$pv_{(i,j,sl)}^{Acq\_0} = \Gamma_{(i,j,sl)} PD_{(i,j,sl)}^{(A)} \exp\left[-\frac{t}{T_{2(i,j,sl)}^{(A)}}\right] \quad \text{(Eq. E3.2–10)}$$

With this protocol, a system of six equations for the six independent components of the diffusion tensor can be generated thus generating the minimum amount of experimental information required for pixel-by-pixel diffusion tensor determination. The results are expressed in terms of the logarithms of the pixel value ratios, specifically:

$$IM_{(i,j,sl)}^{(\lambda)} \equiv \ln\frac{pv_{(i,j,sl)}^{Acq\_\lambda}}{pv_{(i,j,sl)}^{Acq\_0}} = -2b\,\hat{r}^{\lambda^{T}} \cdot \mathbf{D}_{(i,j,sl)} \cdot \hat{r}^{\lambda} \quad \text{(Eq. E3.2–11)}$$

where $2b = \alpha^2 g_0^2$. In this case which is algebraically tractable (P. Basser & Pierpaoli, 1998), the six independent components of the diffusion tensor for each pixel are:

$$\left(D_{xz}^{(eff)}\right)_{(i,j,sl)} = \frac{-IM_{(i,j,sl)}^{(1)} + IM_{(i,j,sl)}^{(2)}}{4b}; \quad \left(D_{yz}^{(eff)}\right)_{(i,j,sl)} = \frac{-IM_{(i,j,sl)}^{(3)} + IM_{(i,j,sl)}^{(4)}}{4b}$$

$$\left(D_{xy}^{(eff)}\right)_{(i,j,sl)} = \frac{-IM_{(i,j,sl)}^{(5)} + IM_{(i,j,sl)}^{(6)}}{4b}$$

$$\left(D_{xx}^{(eff)}\right)_{(i,j,sl)} = \frac{-IM_{(i,j,sl)}^{(1)} - IM_{(i,j,sl)}^{(2)} + IM_{(i,j,sl)}^{(3)} + IM_{(i,j,sl)}^{(4)} - IM_{(i,j,sl)}^{(5)} - IM_{(i,j,sl)}^{(6)}}{4b}$$

$$\left(D_{yy}^{(eff)}\right)_{(i,j,sl)} = \frac{IM_{(i,j,sl)}^{(1)} + IM_{(i,j,sl)}^{(2)} - IM_{(i,j,sl)}^{(3)} - IM_{(i,j,sl)}^{(4)} - IM_{(i,j,sl)}^{(5)} - IM_{(i,j,sl)}^{(6)}}{4b}$$

$$\left(D_{zz}^{(eff)}\right)_{(i,j,sl)} = \frac{-IM_{(i,j,sl)}^{(1)} - IM_{(i,j,sl)}^{(2)} - IM_{(i,j,sl)}^{(3)} - IM_{(i,j,sl)}^{(4)} + IM_{(i,j,sl)}^{(5)} + IM_{(i,j,sl)}^{(6)}}{4b}$$

(Eq. E3.2–12)

There are infinite gradient schemes for DTI; the example above is simple and minimalistic and is used for illustrative purposes only.

This gives the diffusion tensor in the laboratory system of coordinates. By solving the eigenvalue problem at every pixel location, specifically:

$$\mathbf{D}_{(i,j,sl)}\vec{\xi}_{(i,j,sl)} = \lambda_{(i,j,sl)}\vec{\xi}_{(i,j,sl)} \qquad \text{(Eq. E3.2–13)}$$

the diffusion tensor can be expressed in a local coordinate system, which is characterized by the three principal directions of diffusivity or equivalently by the directions of the three eigenvectors,

$$\left\{ \vec{\xi}^{(1)}_{(i,j,sl)}; \quad \vec{\xi}^{(2)}_{(i,j,sl)}; \quad \vec{\xi}^{(3)}_{(i,j,sl)} \right\} \qquad \text{(Eq. E3.2–14)}$$

which are orthogonal to each other.

In such local coordinate system, the diffusion tensor is diagonal and the diagonal elements are the principal directional diffusivities, *i.e.*

$$\mathbf{D}_{(i,j,sl)} \longrightarrow \mathbf{D}^{\text{dia}}_{(i,j,sl)} = \begin{pmatrix} \lambda^{(1)}_{(i,j,sl)} & 0 & 0 \\ 0 & \lambda^{(2)}_{(i,j,sl)} & 0 \\ 0 & 0 & \lambda^{(3)}_{(i,j,sl)} \end{pmatrix} \qquad \text{(Eq. E3.2–15)}$$

Several scalar measures of anisotropic diffusion can be derived from the diagonal DT; among the most useful and widely used are the mean diffusivity and fractional anisotropy, which are respectively given by:

$$\bar{\lambda}_{(i,j,sl)} = \frac{1}{3}\left( \lambda^{(1)}_{(i,j,sl)} + \lambda^{(2)}_{(i,j,sl)} + \lambda^{(3)}_{(i,j,sl)} \right) \qquad \text{(Eq. E3.2–16)}$$

and,

$$\text{FA}_{(i,j,sl)}$$
$$= \sqrt{\frac{2}{3}} \sqrt{\frac{(\lambda^{(1)}_{(i,j,sl)} - \bar{\lambda}_{(i,j,sl)})^2 + (\lambda^{(2)}_{(i,j,sl)} - \bar{\lambda}_{(i,j,sl)})^2 + (\lambda^{(3)}_{(i,j,sl)} - \bar{\lambda}_{(i,j,sl)})^2}{(\lambda^{(1)}_{(i,j,sl)})^2 + (\lambda^{(2)}_{(i,j,sl)})^2 + (\lambda^{(3)}_{(i,j,sl)})^2}}$$
$$\text{(Eq. E3.2–17)}$$

As defined, fractional anisotropy is a parameter that ranges from 0 (minimum anisotropy) to 1 (maximum anisotropy).

### E3.2.ii. *Diffusion qMRI Pulse Sequences*

Most diffusion qMRI pulse sequences are designed around the pulsed field gradient pulses originally described by Stejskal and Tanner (Stejskal &

Tanner, 1965), and these are generically known as PFG diffusion weighted pulse sequences (see Fig. E-3, 4).

*In vivo* PFG diffusion imaging is challenging because of its extreme sensitivity to nondiffusive bulk motions; e.g. respiration, pulsations, flow, peristalsis, and gross patient motion, and because of the concomitant $T_2$-weighting resulting from operating in the medium-long TE regime as needed in order to accommodate the diffusion pulses at useful b-values. The artifactual effects of the nondiffusive motions depend on the body part and on the PFG pulse sequence. Norris (Norris, 2001) reviewed and organized logically by operational principle, techniques for reducing the effects of nondiffusive motions in PFG diffusion MRI. Reviewed techniques include pulse sequences that are very fast thus effectively "freezing" some of the non-diffusional motions, techniques which do not rely on phase encoding for spatial localization, techniques that correct phase errors by means of navigator echoes, and cardiac gated techniques with which scanning during the systolic portion of the cardiac cycle is avoided.

Perhaps the most useful modification to the original PFG spin-echo pulse sequence has been replacing the conventional one-k-line-per-TR readout:

$$\left(\frac{\pi}{2}\right)_x \xrightarrow{\quad PFG_1(\delta) \quad} (\pi)_x \xrightarrow{\quad PFG_2(\delta) \quad} SE\_readout \qquad \text{(Eq. E3.2–18)}$$

with a single-shot EPI readout (Turner *et al.*, 1990), thus leading to the single shot spin-echo EPI sequence,

$$\left(\frac{\pi}{2}\right)_x \xrightarrow{\quad PFG_1(\delta) \quad} (\pi)_x \xrightarrow{\quad PFG_2(\delta) \quad} EPI\_readout \qquad \text{(Eq. E3.2–19)}$$

which is the current workhorse of clinical diffusion MRI. Due to its high data acquisition speed –typically less than 100 ms per full k-space readout — this basic pulse sequence gives accurate and reproducible results, albeit with several limitations and pitfalls (Le Bihan, Poupon, Amadon, & Lethimonnier, 2006). These include 1) low spatial resolution, 2) high vulnerability to geometric distortion artifacts in areas of magnetic field inhomogeneity, 3) residual motion artifact sensitivity to cardiac pulsations, and 4) large chemical shift artifacts. Improvements in spatial resolution become possible with more powerful gradient systems, the use of which also leads to shorter TEs for a given b-value and therefore higher SNR due to less concomitant $T_2$-weighting. Second, because the severity of the geometric distortion artifacts increases as a function of the echo train length of the EPI readout, image quality improvements can be obtained with parallel imaging techniques whereby a reduction in the number of acquired k-lines becomes possible. Third, reduction in residual motion artifacts can be achieved by either using cardiac gating or more efficiently with the use of navigator echoes. Forth, chemical shift artifacts mandate use of a fat suppression prepulse or use of the spatial-spectral excitation pulse.

A fundamentally different technique for PFG diffusion MRI termed line scan diffusion imaging (LSDI) (Gudbjartsson *et al.*, 1996) consists of scanning sequentially parallel columns --or "lines"-- of tissue that are formed at the intersection of two intersecting oblique slices: one slice is defined by a slice selective excitation pulse and the other defined by a slice selective refocusing pulse. Unlike the single-shot EPI technique, whereby a full k-space plane is traversed using a combination of frequency encoding in one direction and of phase encoding in the other, phase encoding is not used at all with LSDI pulse sequences. Accordingly, the LSDI raw data of an image "lives" in an intermediate signal space whereby one dimension corresponds to geometric space and the other to k-space. Hence, only a one-dimensional Fourier transform along the frequency encoding direction is needed for image reconstruction. The sequential collection of this line-by-line datasets makes LSDI pulse sequences largely insensitive to bulk motion artifacts because phase encoding is not used. In addition, LSDI pulse sequences are less vulnerable to geometrical distortion artifacts in areas of magnetic field inhomogeneity relative to EPI sequences. However, the LSDI sequences are four to six times slower than single-shot diffusion-weighted EPI (Kubicki *et al.*, 2004). In a comparative study performed with healthy volunteers, these investigators found that both EPI- and LSDI-derived FA measures are both sufficiently robust but that when higher accuracy is needed, LSDI provides smaller error and smaller inter-subject and inter-session variability than single-shot EPI (Kubicki, *et al.*, 2004).

PFG techniques rely on using strong and balanced gradient pulses for diffusion sensitization and these pulses can induce transient electrical currents in the magnet's inner metallic structures. In turn, these unwanted eddy currents are source to spurious magnetic fields, which can result in imbalance of the gradient pulses as well as distortions of the imaging gradients. Imbalanced diffusion gradient pulses lead to artifactual signal decay by imperfect refocusing and consequently, lead to overestimating the diffusion coefficients. To complicate these matters further, eddy current artifacts are in general position dependent; hence, diffusion qMRI errors can show intricate geometric patterns as well as causing registration imperfections between the differently weighted images. Scanner manufacturers go to great lengths in minimizing eddy currents and this is achieved in great measure by using actively shielded gradient coils that minimize the magnetic fields outside the coils. Eddy currents can be further minimized at the pulse sequence level by using gradient pulses of opposite polarity for generating cancelling eddy currents. One such approach is embodied in the twice-refocused spin-echo (TRSE) pulse sequence, which uses four diffusion gradient pulses and two 180° refocusing pulses as opposed to only two gradient pulses and one 180° in the standard PFG approach. The basic idea is to divide the unipolar gradient pulses into bipolar gradient pairs:

$$\left(\frac{\pi}{2}\right)_x \xrightarrow{\text{PFG}_1(\delta_1)} (\pi)_x \xrightarrow{\text{PFG}_2(\delta_2)} \xrightarrow{\text{PFG}_3(\delta_3)} (\pi)_x \xrightarrow{\text{PFG}_4(\delta_4)} \text{EPI\_readout}$$

(Eq. E3.2–20)

so to generate counteracting eddy currents that effectively minimize the eddy fields presence. Diffusion weighting with the TRSE sequence is therefore accomplished by applying two bipolar diffusion pulses each of which is "divided" by a 180° refocusing pulse. In this implementation (Reese, Heid, Weisskoff, & Wedeen, 2003), the following equations apply,

$$\left.\begin{aligned} \delta_1 + \delta_2 &= \delta_3 + \delta_4 \\ \delta_2 + \delta_3 &= \frac{\text{TE}}{2} \\ \delta_1 + \delta_4 &= \frac{\text{TE}}{2} - t_{\text{pr}} \end{aligned}\right\}$$

(Eq. E3.2–21)

where, $t_{\text{pr}}$ is the sum of the preparation time following the excitation pulse and the readout time preceding the SE.

Compared to standard PFG, the TRSE pulse sequence provides high immunity to eddy current artifacts with minimal scanning efficiency penalties; only the extra duration of the second 180° refocusing pulse. Furthermore, the b-value formula of TRSE diffusion weighted pulse sequences is:

$$b^{\text{Acq-}\lambda} = (2\pi\gamma)^2 g_\lambda^2 (\delta_1 + \delta_2)^2 \left[\Delta - \frac{(\delta_1 + \delta_2)}{3}\right]$$

(Eq. E3.2–22)

where, for simplicity, we have neglected the effects of finite gradient pulse rise times.

A recently described (Holdsworth, Skare, Newbould, & Bammer, 2009) pulse sequence for diffusion tensor MRI incorporates several of the elements discussed above. It begins with a water-selective and slice-selective excitation pulse. Second diffusion encoding ensues using the TRSE approach thus minimizing eddy currents follows it. Thirdly, several k-lines are read with an asymmetric EPI readout for minimizing TE, and finally, a third 180° refocusing pulse that forms a navigator echo used for correcting the k-space data of the just acquired blind for phase errors. With this pulse sequence, investigators obtain diffusion tensor images of high image quality that rival that of standard non-qMRI clinical scans.

Another diffusion encoding gradient scheme has been described whereby oscillating gradients (Does, Parsons, & Gore, 2003; Schachter, Does, Anderson, & Gore, 2000) are used instead of the typical rectangular waveforms with fixed gradient amplitudes. The resulting spin-echo pulse sequence, termed logically oscillating gradient spin-echo (OGSE) essentially implements a succession of diffusion-weighting periods and is useful for probing the diffusion coefficient as function of the diffusion time. This is of particular interest for studying the short to medium diffusion times with associated diffusion lengths from the sub-cellular of a few µm, to the structural tissue scales. The timing diagram of one such OGSE-EPI pulse sequence is very similar to the DW-SE-EPI sequence but differs in the use of sinusoidal waveforms of the diffusion encoding pulses (Does, *et al.*, 2003).

### E3.3. *qMRI of Flow and Displacement*

We consider a moving spin packet with instantaneous position $\vec{x}(t)$ that is initially excited to the transverse state, and we study the subsequent NMR dynamics in the presence of the inhomogeneity field, which is static, plus a time dependent magnetic field gradient. Accordingly, the z-component of the magnetic field is:

$$B_z^{(App)}(\vec{x}, t) = B_0 + \delta B_0(\vec{x}) + \vec{x}(t) \cdot \vec{g}(t) \qquad \text{(Eq. E3.3–1)}$$

Consequently, the transverse BTS equation reads:

$$\frac{\partial m^{(A)}}{\partial t} = -i(\omega_0 + \delta\omega_0(\vec{x}) + 2\pi\gamma\vec{x}(t) \cdot \vec{g}(t))m^{(A)} - \frac{m^{(A)}}{T_2^{(A)}} - \vec{v}^{(A)} \cdot \vec{\nabla}m^{(A)}$$

$$\text{(Eq. E3.3–2)}$$

This equation can be readily integrated resulting in the flollowing analytical solution in the rotating frame of reference:

$$m^{(A)}(\vec{x}, t) = m^{(A)}(\vec{x}, 0) \exp\left(-\left(i\delta\omega_0(\vec{x}) + \frac{1}{T_2^{(A)}}\right)t\right)$$

$$\times \exp\left(-i2\pi\gamma \int_0^t \vec{x}(t') \cdot \vec{g}(t')dt'\right) \qquad \text{(Eq. E3.3–3)}$$

Hence, if the spin packet has a constant position, the resulting phase shift is equal to:

$$\Phi(\vec{x}, t) = \delta\omega_0(\vec{x})t + 2\pi\gamma\vec{x} \cdot \vec{\Lambda}^{(0)} \qquad \text{(Eq. E3.3–4)}$$

where the zero order time moment of the gradient waveform, specifically

$$\vec{\Lambda}^{(0)} \equiv \int_0^t \vec{g}(t')dt' \qquad \text{(Eq. E3.3–5)}$$

We use here the symbol $\vec{\Lambda}^{(n)}$ to denote the $n^{\text{th}}$-order time moment of a gradient waveform and not the commonly used (Pelc, Bernstein, Shimakawa, & Glover, 1991) notation $\vec{M}^{(n)}$, to avoid confusion with the symbol used for the magnetization. Many gradient waveforms used in MRI pulse sequences have null zero order time moment; for example the symmetrically balanced bipolar pulses such that:

$$g(t) = \begin{cases} 0 & t < t_0 \\ +g^{(max)} & t_0 \le t \le t_0 + \delta \\ 0 & t_0 + \delta < t < t_0 + \Delta \\ -g^{(max)} & t_0 + \Delta \le t \le t_0 + \Delta + \delta \\ 0 & t > t_0 + \Delta + \delta \end{cases} \qquad \text{(Eq. E3.3–6)}$$

In such case, the phase shift accrued by an immobile spin packet is:

$$\Phi(\vec{x}, t) = -\delta\omega_0(\vec{x})t \qquad \text{(Eq. E3.3–7)}$$

If on the other hand, a spin packet is moving with an approximately constant velocity, the position vector can be written as a Taylor's series, that is:

$$\vec{x}(t') \approx \vec{x}(t) + \vec{V}^{(A)}(t' - t) + \text{higher order terms} \qquad \text{(Eq. E3.3–8)}$$

and the transverse magnetization is:

$$m^{(A)}(\vec{x}, t) = m^{(A)}(\vec{x}, 0) \exp\left(-\left(i\delta\omega_0(\vec{x}) + \frac{1}{T_2^{(A)}}\right)t\right)$$

$$\times \exp\left(-i2\pi\gamma\vec{x}(t) \cdot \int_0^t \vec{g}(t')dt'\right)$$

$$\times \exp\left(-i2\pi\gamma\vec{v}^{(A)} \cdot \int_0^t \vec{g}(t')(t' - t)dt'\right) \qquad \text{(Eq. E3.3–9)}$$

Hence, the phase shift accrued by a moving spin packet at a time t after the end of a balanced dipolar gradient pulse –*i.e.* with null zero order time moment-- can be written in terms of the first order time moment of the gradient waveform,

$$\Phi^{(\text{flow})}(\vec{x}, t) = -\delta\omega_0(\vec{x})t - 2\pi\gamma\vec{v}^{(A)} \cdot \vec{\Lambda}^{(1)}(t) \qquad \text{(Eq. E3.3–10)}$$

or equivalently,

$$\boxed{\Phi^{(\text{flow})}(\vec{x}, t) = -\delta\omega_0(\vec{x})t - 2\pi\gamma \sum_{\alpha=\text{fe,pe,sl}} v_\alpha^{(A)}\Lambda_\alpha^{(1)}(t) \qquad \text{(Eq. E3.3–11)}}$$

where the sum runs over the three physical directions of the pulse sequence and the first order time moment of the gradient waveform in the $\alpha$-direction has been defined by:

$$\Lambda_\alpha^{(1)}(t) \equiv \int_0^t g_\alpha(t')(t - t')dt' \quad \text{where } \alpha = \text{fe, pe, sl} \qquad \text{(Eq. E3.3–12)}$$

The first order time moment for a symmetric bipolar pulse is:

$$\Lambda_\alpha^{(1)}(t) = G_\alpha\tau^2 \qquad \text{(Eq. E3.3–13)}$$

where $\mathbf{G}_\alpha$ and $\tau$ are respectively the amplitude and duration of the first lobe; the second lobe has same duration but opposite polarity.

    Equations E3.7-10-12 are the theoretical building blocks of qMRI of flow and bulk motion by differential velocity weighting of phase images, also known a quantitative flow by phase contrast. The three components of the velocity can be quantified by acquiring images with the first order moment vector along different directions and keeping its magnitude fixed.

    The equations also these suggest a velocity weighting qCV for phase imaging, specifically for each spatial direction:

$$V_\alpha^{(\text{enc})} = \frac{1}{2\gamma\Lambda_\alpha^{(1)}(\text{TE})} \quad \text{where } \alpha = \text{fe, pe, sl} \qquad \text{(Eq. E3.3–14)}$$

Such encoding velocity --often referred to as VENC-- represents the maximum velocity encoded up to the time of measurement TE. Hence, by generating a minimum of four phase images with judiciously chosen values of $V_\alpha^{(\text{enc})}$ --or equivalently with different first order moments--, mappings of each component of the velocity vector can be calculated. Typically, the four values of the velocity encoding qCV are chosen as $V_\alpha^{(\text{enc})} = \text{VENC}$ for all three directions $\alpha = \text{fe}, \text{pe}, \text{sl}$, plus an additional phase image acquired without velocity encoding, *i.e.* $V_\alpha^{(\text{enc})} = \infty$, resulting in a system of four equations:

$$
\left.
\begin{aligned}
\Phi_{\text{fe}}^{(\text{flow})}(\vec{X}, \text{TE}) &= -\delta\omega_0(\vec{x})\text{TE} - \frac{V_{\text{fe}}^{(A)}}{2\gamma\Lambda_\alpha^{(1)}(\text{TE})} \\[2ex]
\Phi_{\text{pe}}^{(\text{flow})}(\vec{X}, \text{TE}) &= -\delta\omega_0(\vec{x})\text{TE} - \frac{V_{\text{pe}}^{(A)}}{2\gamma\Lambda_\alpha^{(1)}(\text{TE})} \\[2ex]
\Phi_{\text{sl}}^{(\text{flow})}(\vec{X}, \text{TE}) &= -\delta\omega_0(\vec{x})\text{TE} - \frac{V_{\text{sl}}^{(A)}}{2\gamma\Lambda_\alpha^{(1)}(\text{TE})}
\end{aligned}
\right\}
\qquad \text{(Eq. E3.3--15)}
$$

and independently,

$$
\Phi^{(\text{ref})}(\vec{X}, \text{TE}) = -\delta\omega_0(\vec{X})\text{TE} \qquad \text{(Eq. E3.3--16)}
$$

With these equations, each component of the velocity vector can be calculated in terms of the measured phase differences at each pixel using the following three formulas:

$$
\boxed{
V_\alpha^{(A)} = \frac{\Phi_\alpha^{(\text{flow})}(\vec{X}, \text{TE}) - \Phi^{(\text{ref})}(\vec{X}, \text{TE})}{2\pi\gamma\Lambda_\alpha^{(1)}(\text{TE})} \quad \text{where } \alpha = \text{fe}, \text{pe}, \text{sl}
}
$$

$$\text{(Eq. E3.3--17)}$$

When implemented with a pixel wise algorithm, maps of each velocity component can be generated. Often one is interested only in the speed and the magnitude of the flow velocity is:

$$
\boxed{
|\vec{V}^{(A)}| = \frac{1}{2\pi\gamma\Lambda_\alpha^{(1)}(\text{TE})} \sqrt{\sum_{\alpha=\text{fe},\text{pe},\text{sl}} (\Phi_\alpha^{(\text{flow})}(\vec{X}, \text{TE}) - \Phi^{(\text{ref})}(\vec{X}, \text{TE}))^2}
}
$$

$$\text{(Eq. E3.3--18)}$$

The last two equations above are the primary building blocks for flow and motion qMRI algorithms by the method of phase contrast (PC) that was originally described by Moran (Moran, 1982). Several encoding schemes (Pelc, *et al.*, 1991) are possible: the six-point method uses three pairs of images that include velocity encoded images along the three orthogonal directions plus three reference phase images without flow encoding. With the simple four-point method, only one reference phase image is acquired and used for calculating each velocity component. Other velocity encoding strategies that involve applying velocity encoding gradients in two directions at a time have also been studied (Pelc, *et al.*, 1991).

For over twenty years, phase contrast imaging has been the method of choice for quantifying blood flow with MRI. This is however, an area of active research and newer methods have been described in the literature (Nielsen & Nayak, 2009; Zuo, Walsh, Deutsch, & Twieg, 2006).

Phase contrast velocity encoding can be used also for quantifying the displacement of tissue during a certain time interval. Such applications include the study of deformations of the myocardium during the cardiac cycle and the study of tissue elasticity as studied with an external source of displacement. In these applications, the object of interest is not the average velocity over a short period of time but the actual tissue displacement over longer time interval of about 100 ms. For such applications, a displacement encoding *via* stimulated echoes (DENSE) has been described in the literature (Aletras, Ding, Balaban, & Wen, 1999). It offers the ability to extract myocardial motion data at high spatial density over segments of the cardiac cycle. Other techniques based on cardiac tagging offer lower spatial densities in comparison because these are rely on tagging the magnitude images with a grid of signal voids during rf excitation and subsequently tracking the lower spatial density tags as function of time (Axel & Dougherty, 1989a, 1989b; Zerhouni, Parish, Rogers, Yang, & Shapiro, 1988). The DENSE technique for displacement mapping is based on a special stimulated echo pulse sequence with added displacement encoding gradient pulses applied along a desired direction during the time interval between the first two 90° pulses and the matching lobe applied between the third 90° pulse and the signal readout. With the described implementation (Aletras, *et al.*, 1999), imaging is performed with slice selection during the third rf pulse followed by sequential k-space sampling at one k-line per excitation. Maps of the displacement along each direction are computed as the difference of phase images:

$$\Delta \varphi_\alpha = 2\pi\gamma(G_\alpha - G_\alpha^*)\Delta X_\alpha \qquad \text{(Eq. E3.3–19)}$$

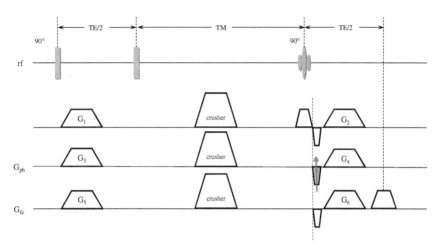

Fig. E-5. Timing diagram of the DENSE pulse sequence (adapted from (Aletras, *et al.*, 1999)). Spin phase wrapping occurs through the application of gradient pulses $G_1$, $G_3$, and $G_5$. The corresponding gradient pulses $G_2$, $G_4$, and $G_6$ totally unwrap the phase for static tissue. Spins that have moved during the TM period will acquire a phase that is proportional to the displacement (see equation above). Crusher gradient pulses are applied during the mixing period TM.

that are acquired with two different gradient amplitudes $G_\alpha$ and $G_\alpha^*$ for the displacement encoding gradient pulses along each direction.

### E3.4. qMRI of the Longitudinal Relaxation Time ($T_1$)

qMRI offers several possibilities for mapping the longitudinal relaxation time. As mentioned above, the basic unifying principle is to acquire several image datasets with varying degrees of weighting to the qMRI parameter targeted for mapping, in this case the longitudinal relaxation time. Hence rewriting the general mapping equation:

$$\text{qMRI\_par\_X}_{(i,j,sl)}^{(A)} = \Re(\text{pv}_{(i,j,sl)}^{\text{Acq\_1}}, \ldots, \text{pv}_{(i,j,sl)}^{\text{Acq\_n}}, \text{qCV\_X}(1, 2, \ldots, n))$$

$$(\text{Eq. E3.4--1})$$

and applying it to the longitudinal relaxation time, we have:

$$T_{1,(i,j,sl)}^{(A)} = \Re(\text{pv}_{(i,j,sl)}^{\text{Acq\_1}}, \ldots, \text{pv}_{(i,j,sl)}^{\text{Acq\_n}}, \text{qCV\_T1}(1, 2, \ldots, n)) \quad (\text{Eq. E3.4--2})$$

where the qCVs most often used for $T_1$ quantification are the inversion time, the repetition time, and the flip (nutation) angle for inversion recovery, saturation recovery, and variable nutation angle techniques respectively. In other words:

$$qCV\_T1(1,2,\ldots,n) = \begin{cases} TI(1,2,\ldots,n) & \rightarrow \quad \text{Inversion recovery} \\ TR(1,2,\ldots,n) & \rightarrow \quad \text{Saturation recovery} \\ FA(1,2,\ldots,n) & \rightarrow \quad \text{Variable nutation angle} \end{cases}$$
$$\text{(Eq. E3.4-3)}$$

In the following sections, we derive the specific form for the algorithm function $\Re$ that is applicable in each case. To this end, we need the pixel value equation that describes the directly acquired images as a function of the qCV relevant to the quantification strategy.

### E3.4.i. *Inversion Recovery Techniques: qCV = TI*

As the name implies, inversion recovery $T_1$ mapping techniques are variations of the inversion recovery experiment whereby measurements are performed at n TI times as measured relative to the time at which the inversion pulse is applied.

In the most rudimentary and slowest implementation, one k-line of one slice is measured for each inversion pulse and the pixel value equation for such a conventional inversion recovery spin-echo (IR-SE) sequence is:

$$pv_{(i,j,sl)}^{Acq\_\lambda} = \Gamma_{(i,j,sl)} PD_{(i,j,sl)}^{(A)} \left( 1 - (1 - \cos(FA)) \exp\left( -\frac{TI^{Acq\_\lambda}}{T_{1,(i,j,sl)}^{(A)}} \right) \right.$$

$$\left. + \exp\left( -\frac{TR}{T_{1,(i,j,sl)}^{(A)}} \right) \right) \exp\left( -\frac{TE}{T_{2,(i,j,sl)}^{(A)}} \right) \qquad \text{(Eq. E3.4-4)}$$

This equation includes a magnetization saturation factor to include the case of pulse sequences performed at a finite TR such that not all tissues can regain the full equilibrium longitudinal magnetization in every pulse sequence cycle. In addition, in this general equation, the flip angle FA of the inversion recovery pulse has been kept explicit since deviations from its nominal 180° value is the leading cause for errors in the $T_1$ estimation, particularly at higher field strengths.

Because the partial saturation factor complicates the mathematics of algorithms for $T_1$ quantification, IR pulse sequences are most often implemented in the long TR or fully unsaturated regime where the more tractable pixel value equation applies:

$$\mathrm{pv}^{\mathrm{Acq\text{-}\lambda}}_{(i,j,sl)} \cong \Gamma_{(i,j,sl)} \mathrm{PD}^{(A)}_{(i,j,sl)} \left( 1 - (1 - \cos(\mathrm{FA})) \exp\left( -\frac{\mathrm{TI}^{\mathrm{Acq\text{-}\lambda}}}{\mathrm{T}^{(A)}_{1,(i,j,sl)}} \right) \right)$$

$$\times \exp\left( -\frac{\mathrm{TE}}{\mathrm{T}^{(A)}_{2,(i,j,sl)}} \right) \qquad\qquad \text{(Eq. E3.4–5)}$$

This pixel value equation is of the general form:

$$\mathrm{pv}^{\mathrm{Acq\text{-}\lambda}}_{(i,j,sl)} = \mathrm{A}_{(i,j,sl)} - \mathrm{B}_{(i,j,sl)} \exp\left( -\frac{\mathrm{TI}^{\mathrm{Acq\text{-}\lambda}}}{\mathrm{T}^{(A)}_{1,(i,j,sl)}} \right) \qquad \text{(Eq. E3.4–6)}$$

where we have defined A and B by:

$$\mathrm{A}_{(i,j,sl)} \equiv \Gamma_{(i,j,sl)} \mathrm{PD}^{(A)}_{(i,j,sl)} \exp\left( -\frac{\mathrm{TE}}{\mathrm{T}^{(A)}_{2,(i,j,sl)}} \right) \qquad \text{(Eq. E3.4–7)}$$

and,

$$\mathrm{B}_{(i,j,sl)} \equiv \Gamma_{(i,j,sl)} \mathrm{PD}^{(A)}_{(i,j,sl)} (1 - \cos(\mathrm{FA})) \exp\left( -\frac{\mathrm{TE}}{\mathrm{T}^{(A)}_{2,(i,j,sl)}} \right) \quad \text{(Eq. E3.4–8)}$$

Therefore, $\mathrm{T}^{(A)}_1$ can be mapped using Eq. E3.4-6 with a three-parameter fitting routine.

Alternatively, for two timepoint techniques, *i.e.* two TI times, a semi-analytical equation based on the root finding routine can be used, specifically:

$$\mathrm{T}^{(A)}_{1,(i,j,sl)}$$

$$= \mathrm{Root}\left( \mathrm{T}^{(A)}_{1,(i,j,sl)} + \frac{\mathrm{TI}^{\mathrm{Acq\text{-}1}}}{\ln\left[ \frac{1}{2}\left( 1 - \left( \frac{\mathrm{pv}^{\mathrm{Acq\text{-}1}}_{(i,j,sl)}}{\mathrm{pv}^{\mathrm{Acq\text{-}2}}_{(i,j,sl)}} \right)\left( 1 - (1 - \cos(\mathrm{FA})) \times \exp\left( -\frac{\mathrm{TI}^{\mathrm{Acq\text{-}2}}}{\mathrm{T}^{(A)}_{1,(i,j,sl)}} \right) \right) \right) \right]}, \tau 1 \right)$$

$$\text{(Eq. E3.4–9)}$$

where $\tau$, is an initial guess value of the solution.

The fully unsaturated conventional IR-SE pulse sequence leads to scan times that are very long thus precluding its implementation in clinical practice. For example, acquiring one $256 \times 256$ image with a conventional readout –*i.e.* one k-line per TR-- inversion recovery pulse sequence with $TR = 10\,s$ and $n = 10$ TI values, would take in excess of 7 hours. In response to this problem, the aim has been to develop faster IR pulse sequences. Numerous such fast-IR $T_1$-qMRI pulse sequences have been described in the scientific literature (Bluml, Schad, Stepanow, & Lorenz, 1993; Deichmann, Hahn, & Haase, 1999; P. Gowland & Mansfield, 1993; P. A. Gowland & Leach, 1992; Haase, 1990; Henderson, McKinnon, Lee, & Rutt, 1999; Kay & Henkelman, 1991; Scheffler & Hennig, 2001; Schmitt *et al.*, 2004; Steinhoff, Zaitsev, Zilles, & Shah, 2001) and this is still an active area of research. The central problem is how to acquire a given imaging volume at several (n) time points post application of an inversion pulse thus generating the directly acquired datasets:

$$\{pv_{(i,j,sl)}^{Acq-\lambda}\} \quad \text{for} \begin{cases} \lambda = 1, \ldots, n \\ \text{and} \\ (i, j, sl) = (1, 1, 1) \cdots (N_x, N_y, N_s) \end{cases} \qquad \text{(Eq. E3.4–10)}$$

that are needed for mapping $T_1$ over the targeted imaging volume. As additional requirement, the resulting scan time should be about a few minutes. The question of how many time points *versus* spatial resolution, spatial coverage, and signal to noise is not simple and depends on many factors ultimately riding on hardware considerations. In principle, if the return to equilibrium is monoexponential, then only two inversion times are necessary for mapping $T_1$ and therefore most of the scan time should be "invested" in the geometric aspects of the scan, *e.g.* spatial resolution, or spatial coverage, or a tradeoff mixture of both. Conversely, availability of data at more than two TI time points can be used for improving the accuracy of the data fits, as well as for studying potential multiexponential recovery as it may occur in some tissues. For typical clinical applications, one needs high spatial resolution, comprehensive spatial coverage, and adequate temporal resolutions along the recovery curve.

Two types of fast IR qMRI techniques can be identified: the ones that interrogate the k-space data sequentially at each one time TI along the recovery curve *versus* the multiple TI techniques that stem from the Look Locker concept that was first proposed by Look and Locker for NMR (Look & Locker, 1970). IR-Look-Locker (LL) techniques interrogate the longitudinal magnetization using a series evenly spaced timepoints post-inversion, using small flip angle excitations and fast readouts. The net result is a significant reduction in acquisition time compared to standard IR techniques. With LL imaging pulse sequences, one may sample all of k-space at each time increment using either single shot echo planar imaging (EPI) (P. Gowland & Mansfield, 1993) or fast low-angle shot (FLASH) (Haase, 1990). Another possible approach is the method of $T_1$ mapping by sampling a single line of k-space multiple times following an inversion pulse is known as T One by Multiple Read Out Pulses (TOMROP) (Brix, Schad, Deimling, & Lorenz, 1990).

The succession of excitation pulses in LL pulse sequences affect the recovery of the longitudinal magnetization and lead to the estimation of $T_1$ indirectly *via* a related parameter (Kay & Henkelman, 1991) known as the effective longitudinal relaxation time, specifically:

$$\frac{1}{T^*_{1,(i,j,sl)}} = \frac{1}{T^{(A)}_{1,(i,j,sl)}} - \left(\frac{1}{\tau}\right) \ln(\cos(\alpha)) \qquad \text{(Eq. E3.4--11)}$$

where $\alpha$ and $\tau$ are the excitation flip angle and the readout sampling time respectively. Hence, the effective or measured $T^*_1$ is longer than the true $T^{(A)}_1$ and furthermore it is sensitive to imperfections of the readout flip angle alpha. An additional consequence of using the LL train of measurements is a reduced apparent equilibrium longitudinal magnetization $M^{(eq)*}_z$ and as discussed below, this phenomenon provides an often-used alternative methodology for mapping $T_1$ with reduced vulnerability to imperfections in the rf alpha pulses (Deichmann, *et al.*, 1999; Steinhoff, *et al.*, 2001).

For IR-LL type pulse sequences, it can be shown (Deichmann, *et al.*, 1999) that the pixel value equations that describe each of the n differentially $T_1$-weighted images are:

$$pv^{Acq\_\lambda}_{(i,j,sl)} = A_{(i,j,sl)} - B_{(i,j,sl)} \exp\left(-\frac{TI^{Acq\_\lambda}}{T^*_{1,(i,j,sl)}}\right) \qquad \text{(Eq. E3.4--12)}$$

where the factors A and B are respectively given by:

$$A_{(i,j,sl)} = \xi\Gamma_{(i,j,sl)}PD^{(A)}_{(i,j,sl)} \exp\left(-\frac{TE}{T^{(A)*}_{2,(i,j,sl)}}\right) \sin(\alpha) \qquad \text{(Eq. E3.4--13)}$$

and

$$B_{(i,j,sl)} = (1 + \xi)\Gamma_{(i,j,sl)}PD^{(A)}_{(i,j,sl)} \exp\left(-\frac{TE}{T^{(A)*}_{2,(i,j,sl)}}\right) \sin(\alpha) \qquad \text{(Eq. E3.4--14)}$$

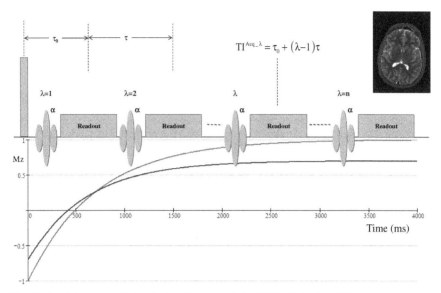

Fig. E-6.  Schematic of the timing diagram of the IR-Look-Locker (LL) pulse sequence (top) and the simulated longitudinal magnetization recovery of two tissues with different $T_1$s. The insert image is a typical $T_1$ map.

and

$$\xi = \frac{M_z^{(eq)*}}{M_z^{(eq)}} \qquad (\text{Eq. E3.4–15})$$

Hence, the effective longitudinal relaxation time can be mapped using a three-parameter fit of Eq. E3.4-12 as function of $TI^{\text{Acq-}\lambda} = \tau_0 + (\lambda - 1)\tau$, and then knowing $T_1^*$ at each pixel location, proceed to map $T_1$ using Eq. E3.4-11. The main drawback of this approach is that $T_1$ accuracy can be compromised by errors in the flip angle. An alternative approach (Deichmann, *et al.*, 1999; Steinhoff, *et al.*, 2001) is to use the equation:

$$T_{1,(i,j,sl)}^{(A)} = T_{1,(i,j,sl)}^* \left( \frac{B_{(i,j,sl)}}{A_{(i,j,sl)}} - 1 \right) \qquad (\text{Eq. E3.4–16})$$

which can be derived by assuming $\tau \ll T_1^*$. In this manner, $T_1$ can be mapped in a manner that does not use Eq. E3.4-11 and is therefore less prone to inaccuracies secondary to imperfections in the rf pulses.

In summary, $T_1$ qMRI with LL-IR techniques can be accomplished with pixel-wise three-parameter fit of images acquired at different inversion times and therefore with different levels of $T_1$-weighting. These images can be acquired in single shot mode, typically with a single shot EPI readout, or with multiple shots in which case banded k-space results. For multislice applications (Steinhoff, *et al.*, 2001), the LL-IR pulse sequence should have loops for k-lines, slice position, and TI time points.

### E3.4.ii. *Saturation Recovery Techniques:* $qCV = TR$

In its most basic implementation, the saturation recovery (SR) pulse sequence consists of a series of 90° excitation pulses, each one followed by a conventional gradient readout thus reading one k-line per TR cycle. For $T_1$ mapping, several such SR sequences are applied in succession, each time using a different value of TR, which is the qCV of this qMRI approach. With SR-GE sequences, there is no need for waiting for full recovery of the longitudinal magnetization between repeated experiments, which makes the IR sequence so time consuming. Therefore, the SR sequences are also referred to as a partial saturation (PS) sequences. The SR sequence is attractive for $T_1$ qMRI because it has a high duty cycle for data sampling. The SR sequence has a factor of two less contrast range than the IR sequence, but this is often overcome by improved signal-to-noise achieved through more frequent sampling.

The pixel value equation for a conventional SR-GE pulse sequence is:

$$pv_{(i,j,sl)}^{Acq\_\lambda} \cong \Gamma_{(i,j,sl)} PD_{(i,j,sl)}^{(A)} \left( 1 - \exp\left( -\frac{TR^{Acq\_\lambda}}{T_{1,(i,j,sl)}^{(A)}} \right) \right) \exp\left( -\frac{TE}{T_{2,(i,j,sl)}^{(A)*}} \right)$$

(Eq. E3.4–17)

which can be written in the familiar form

$$pv_{(i,j,sl)}^{Acq\_\lambda} \cong A_{(i,j,sl)} \left( 1 - \exp\left( -\frac{TR^{Acq\_\lambda}}{T_{1,(i,j,sl)(A)}^{(A)}} \right) \right)$$

(Eq. E3.4–18)

thus permitting $T_1$ mapping *via* two-parameter curve fitting.

Alternatively, for two point techniques, *i.e.* for two TR values, $T_1$ can be mapped using the simple root-finding algorithm following the equation:

$$T_{1,(i,j,sl)}^{(A)}$$

$$= Root \left( T_{1,(i,j,sl)}^{(A)} + \frac{TR^{Acq\_1}}{\ln\left( 1 - \left( \frac{pv_{(i,j,sl)}^{Acq\_1}}{pv_{(i,j,sl)}^{Acq\_2}} \right) \left( 1 - \exp\left( -\frac{TR^{Acq\_2}}{T_{1,(i,j,sl)}} \right) \right) \right)}, \tau 1 \right)$$

(Eq. E3.4–19)

Furthermore, we note that if one of these acquisitions is performed with very long TR, albeit at the expense of scan time, then $T_1$ can be mapped using a very simple analytic expression:

$$T^{(A)}_{1,(i,j,sl)} = \frac{-TR^{Acq\_1}}{\ln\left(1 - \left(\frac{pv^{Acq\_1}_{(i,j,sl)}}{pv^{Acq\_2}_{(i,j,sl)}}\right)\right)} \qquad \text{(Eq. E3.4–20)}$$

as derived from Eq. E.3.4-19 in the infinite TR limit.

### E3.4.iii. *Variable Nutation Angle Techniques: qCV = FA*

The variable nutation angle method, also known as the limited flip angle method (Wang, Riederer, & Lee, 1987), was originally adapted for MRI using a gradient-echo pulse sequence generates maps of $T_1$ with an accuracy similar to that achieved by the IR and SR techniques, but with a significant decrease in acquisition time. In a more recent implementation investigators (Cheng & Wright, 2006; Deoni, 2007; Deoni, Rutt, & Peters, 2003) used sequences that involve establishing a spoiled steady state followed by the collection of spoiled gradient-echo (SPGR) images over a range of flip angles. This generates a signal curve for every pixel that depends on $T_1$. Each signal curve is easily linearized thus allowing $T_1$ mapping.

Starting with the SPGR pixel value equation, specifically:

$$pv^{Acq\_\lambda}_{(i,j,sl)} \cong \Gamma_{(i,j,sl)} PD^{(A)}_{(i,j,sl)} \frac{\left(1 - \exp\left(-\frac{TR}{T^{(A)}_{1,(i,j,sl)}}\right)\right) \sin(FA^{Acq\_\lambda})}{\left(1 - \cos(FA^{Acq\_\lambda}) \exp\left(-\frac{TR}{T^{(A)}_{1,(i,j,sl)}}\right)\right)}$$

$$\times \exp\left(-\frac{TE}{T^{(A)*}_{2,(i,j,sl)}}\right) \qquad \text{(Eq. E3.4–21)}$$

and rewriting it in a manner that is suitable for fitting experimental data as function of flip angle:

$$\frac{pv^{Acq\_\lambda}_{(i,j,sl)}}{\sin(FA)^{Acq\_\lambda}} = \frac{pv^{Acq\_\lambda}_{(i,j,sl)}}{tg(FA^{Acq\_\lambda})} \exp\left(-\frac{TR}{T_{1,(i,j,sl)}}\right)$$

$$+ \Gamma_{(i,j,sl)} PD^{(A)}_{(i,j,sl)} \left(1 - \exp\left(-\frac{TR}{T^{(A)}_{1,(i,j,sl)}}\right)\right) \exp\left(-\frac{TE}{T^{(A)}_{2,(i,j,sl)}}\right)$$

$$\text{(Eq. E3.4–22)}$$

we note that this equation is of the general linear form:

$$Y_{(i,j,sl)} = m_{(i,j,sl)} X_{(i,j,sl)} + b_{(i,j,sl)} \qquad \text{(Eq. E3.4–23)}$$

in which X and Y are parameterized as:

$$Y_{(i,j,sl)} = \frac{pv_{(i,j,sl)}^{Acq\_\lambda}}{\sin(FA^{Acq\_\lambda})} \qquad \text{(Eq. E3.4–24)}$$

and

$$X_{(i,j,sl)} = \frac{pv_{(i,j,sl)}^{Acq\_\lambda}}{tg(FA^{Acq\_\lambda})} \qquad \text{(Eq. E3.4–25)}$$

Consequently, linear regression allows calculating the slope, specifically:

$$m_{(i,j,sl)} = \exp\left(-\frac{TR}{T_{1,(i,j,sl)}^{(A)}}\right) \qquad \text{(Eq. E3.4–26)}$$

and the intercept which is given by:

$$b_{(i,j,sl)} = \Gamma_{(i,j,sl)} PD_{(i,j,sl)}^{(A)}\left(1 - \exp\left(-\frac{TR}{T_{1,(i,j,sl)}^{(A)}}\right)\right) \exp\left(-\frac{TE}{T_{2,(i,j,sl)}^{(A)*}}\right)$$

$$\text{(Eq. E3.4–27)}$$

From the slope, we can map $T_1$ using the equation:

$$T_{1,(i,j,sl)}^{(A)} = \frac{-TR}{\ln(m_{(i,j,sl)})} \qquad \text{(Eq. E3.4–28)}$$

One should note that the slope and the intercept are both functions of $T_1$. This methodology is useful for data acquired with n > 2 different flip angles; alternatively for two point techniques, *i.e.* for two values of the flip angle, we can derive an analytical formula for mapping $T_1$:

$$T_{1,(i,j,sl)}^{(A)} = \frac{-TR}{\ln\left(\dfrac{\sin(FA^{Acq\_1}) - \left(\dfrac{pv_{(i,j,sl)}^{Acq\_1}}{pv_{(i,j,sl)}^{Acq\_2}}\right)\sin(FA^{Acq\_2})}{\left(\sin(FA^{Acq\_1})\cos(FA^{Acq\_2}) - \cos(FA^{Acq\_1}) \times \sin(FA^{Acq\_2})\left(\dfrac{pv_{(i,j,sl)}^{Acq\_1}}{pv_{(i,j,sl)}^{Acq\_2}}\right)\right)}\right)}$$

$$\text{(Eq. E3.4–29)}$$

In summary, the variable nutation angle technique involves using an SPGR pulse sequence and applying a number of times, keeping all scanner settings unchanged. The directly acquired images can be processed using a pixel-by-pixel linear regression algorithm (n > 2), or a simple analytic formula for two point datasets. For high $T_1$ accuracy, stability, and consistency, this method requires complete spoiling of transverse magnetization. A recent study (Preibisch & Deichmann, 2009) finds that the optimum values of the incremental phase used for the rf spoiling for accuracy and stability may not coincide. These investigators propose using an rf spoiling phase of $50°$ that gives stable results at the expense of overestimating $T_1$ and developed an algorithm for correcting the $T_1$ estimates.

### E3.4.iv. *Other $T_1$ qMRI Techniques*

A very different $T_1$ mapping technique, termed fast phase acquisition of composite echoes (FastPACE) uses the phase information of spin-echoes (SEs) and stimulated echoes (STEs) acquired with a single pulse sequence (Ropele, Stollberger, Ebner, & Fazekas, 1999; Ropele, Stollberger, Kapeller, Hartung, & Fazekas, 1999). The original FastPACE sequence consisted of two rf pulses $\alpha_1$ and $\alpha_2$ per phase-encoding step with (typical) flip angles of $45°$ and $90°$ respectively. After the second rf pulse, a composite echo was sampled around the echo time (TE). Assuming full spoiling of higher-order coherences, the generated composite echo is the superposition of an SE and a phase-shifted STE. The SE is formed by the preceding two rf pulses, and the STE is generated by the two rf pulses from the previous phase encoding step and the last rf pulse. Using phase subtraction of two phase-alternated acquisitions, the unknown reference phase was removed, and the following $T_1$-dependent phase shift:

$$\Phi_{(i,j,sl)}^{(T1)} = \arctan\left(\frac{pv_{(i,j,sl)}^{STE}}{pv_{(i,j,sl)}^{SE}}\right) \qquad \text{(Eq. E3.4–30)}$$

could be mapped from the SE and STE pixel value ratios and with this, $T_1$ can be mapped with the simple analytic equation:

$$T_{1,(i,j,sl)}^{(A)} = \frac{TR}{\ln\left(\frac{\cos(\alpha_1)}{tg(\Phi_{(i,j,sl)}^{T1})}\right)} \qquad \text{(Eq. E3.4–31)}$$

As pointed out by this group of researchers in a subsequent paper (Ropele, Bammer, Stollberger, & Fazekas, 2001), "the restrictions of this method arise from the fact that $T_1$ is calculated from a phase image that requires full k-space sampling. Moreover, the SE and STE exhibit a different susceptibility for motion. The simultaneous acquisition of these differently-motion-sensitized signals may not only cause image degradations but may also cause errors in $T_1$ calculation." In response to these technique weaknesses, the investigators develop an improved technique whereby $T_1$ is mapped with the formula:

$$T_{1,(i,j,sl)}^{(A)} = \frac{TR - 0.5\,TE}{\ln\left(\left|\frac{pv_{(i,j,sl)}^{STE}}{pv_{(i,j,sl)}^{SE}}\right|\right)} \qquad \text{(Eq. E3.4–32)}$$

This does not use phase information and is therefore less sensitive to motion degradation. The basic idea behind this modified approach (Ropele, *et al.*, 2001) was to split the SE-STE composite echo into two separately sampled echoes thus avoiding the need for using the phase information of the original SE-STE composite echo.

### E3.5. qMRI of the Transverse Relaxation Time ($T_2$)

E3.5.i. *Multi-SE (CPMG) Techniques: qCV = TE*

A straightforward and intuitive approach for mapping $T_2$ is to acquire a series of spin-echo images with increasing levels of $T_2$-weighting. This is commonly accomplished with a multi spin-echo pulse sequence. The multi-SE NMR pulse sequence that minimizes errors secondary to imperfections in the flip angle of the refocusing 180° pulses has a phase shift of 90° between the excitation pulse –which is set to rotate the magnetization around the x-axis– and each of the refocusing pulses, which are set to rotate the magnetization around the y-axis. This implementation of the multi-SE pulse sequence is referred to as the Carr-Purcell-Meiboom-Gill (CPMG) pulse sequence. Quoting the original paper (Meiboom & Gill, 1958): "With this modification an amplitude deviation of the 180° pulses will not be cumulative in its effect: if the pulses are for instance somewhat less than 180°, the first pulse will leave the polarization vector above the *xy* plane, but the polarization will be returned to this plane on the next pulse." It has been shown however (Poon & Henkelman, 1992) that CPMG imaging pulse sequences are prone to artifacts and that use spoiler gradients applied before and after each refocusing pulse is the most effective technique for artifact suppression. Unfortunately, the use of spoiler gradients renders ineffective radio-frequency phase schemes such as the Meiboom-Gill modification to the Carr-Purcell sequence used to compensate experimental imperfections (Poon & Henkelman, 1992). Hence, the use of the CPMG denomination for these $T_2$ qMRI pulse sequences may not carry the full meaning intended; we will therefore refer to these simply as multi-SE pulse sequences.

With multi-SE pulse sequences, for every position of the imaging grid, n pixel values are generated and $T_2$ can be mapped with a general mapping equation of the form:

$$T_{2,(i,j,sl)}^{(A)} = \Re(pv_{(i,j,sl)}^{Acq\_1}, \ldots, pv_{(i,j,sl)}^{Acq\_n}, qCV\_T2(1, 2, \ldots, n)) \qquad \text{(Eq. E3.5–1)}$$

in which, the quantitative control variable is the echo time, *i.e.*:

$$qCV\_T2(\lambda) = TE_\lambda = \lambda\,ES \qquad \text{(Eq. E3.5–2)}$$

and ES is the inter-echo spacing.

To a first approximation, for a multi-SE pulse sequence the resulting complex pixel values of each echo are given by:

$$pv_{(i,j,sl)}^{Acq\_\lambda} = A_{(i,j,sl)} \exp\left(-\frac{TE_\lambda}{T_{2,(i,j,sl)}^{(A)}}\right) \exp(i\phi_{\lambda(i,j,sl)}) \qquad \text{(Eq. E3.5–3)}$$

where $\phi_\lambda$ denote the individual phases of each echo. With this model, the pixel value equation leads to a standard linear regression problem for the logarithms of the pixel value moduli *versus* TE, based on the following equation:

$$\ln(|pv_{(i,j,sl)}^{Acq\_\lambda}|) = \ln(A_{(i,j,sl)}) - \frac{TE_\lambda}{T_{2(i,j,sl)}^{(A)}} \qquad \text{(Eq. E3.5–4)}$$

Accordingly, the $T_2$ values for each pixel can be obtained from the inverse slopes of these semi-logarithmic linear square fits.

With the idealized formalism above, the mathematics is simple because:

1. The effects of noise have been neglected.
2. The static main magnetic field is assumed perfectly homogeneous.
3. The rf pulses are assumed perfect for all voxels as well as inside each voxel.
4. The effects of inter-slice crosstalk are negligible.
5. The diffusion attenuation effects have been neglected.

$T_2$ qMRI with multi-SE imaging pulse sequences is still an active area of research. In the following, we review the scientific literature and present key results using the notation of this book.

First, in order to study $T_2$ decay in the presence of noise we begin by writing the modified pixel value equation:

$$\text{pv}^{\text{Acq-}\lambda}_{(i,j,sl)} = A_{(i,j,sl)} \exp\left(-\frac{TE_\lambda}{T^{(A)}_{2(i,j,sl)}}\right) \exp(i\phi_{\lambda(i,j,sl)}) + N_{\lambda(i,j,sl)} \exp(i\theta_{\lambda(i,j,sl)})$$

(Eq. E3.5–5)

where the noise term, which is additive and complex-valued, has been written in polar representation. This expression contains to two random variables specifically the noise magnitude N and a phase $\theta$ that have the following associated probability densities (Raya *et al.*, 2010):

$$P_N(n,\sigma) = \frac{n}{\sigma^2} \exp\left(-\frac{n^2}{2\sigma^2}\right)$$

(Eq. E3.5–6)

for the amplitude noise and,

$$P_\theta(\theta) = \frac{1}{2\pi}$$

(Eq. E3.5–7)

for the phase noise.

The method of $T_2$ quantification from quadratic power images (Miller & Joseph, 1993) stems from writing the magnitude square of the pixel values, specifically:

$$|\text{pv}^{\text{Acq-}\lambda}_{(i,j,sl)}|^2 = A^2_{(i,j,sl)} \exp\left(-\frac{TE_\lambda}{T^{(A)}_{2(i,j,sl)}/2}\right) + |N_{(i,j,sl)}|^2 + 2A_{(i,j,sl)}N_{(i,j,sl)}$$

$$\times \exp\left(-\frac{TE_\lambda}{T^{(A)}_{2(i,j,sl)}}\right) \cos(\phi_{\lambda(i,j,sl)} - \theta_{\lambda(i,j,sl)}) \quad \text{(Eq. E3.5–8)}$$

and by noting that the spatial average of last term --*i.e.* the signal-noise cross term-- over a sufficiently large region-of-interest, approaches zero. Hence, assuming an ROI over a homogeneous region of the image, we have:

$$\langle|\text{pv}^{\text{Acq-}\lambda}_{(i,j,sl)}|^2\rangle_{\text{ROI}} \cong \langle A^2_{(i,j,sl)}\rangle_{\text{ROI}} \exp\left(-\frac{TE_\lambda}{\langle T^{(A)}_{2,(i,j,sl)}\rangle_{\text{ROI}}/2}\right) + \langle|N_{(i,j,sl)}|^2\rangle_{\text{ROI}}$$

(Eq. E3.5–9)

This approximate equation can be used in a fitting procedure to an exponential function summed with a positive constant, which leads to a measure of the average relaxation time divided by two. Accordingly, the $T_2$ quantification method using power images and neglecting the cross term is strictly valid at the spatial scale of ROIs only. Nonetheless, Raya and coworkers (Raya, *et al.*, 2010) extended this interpretation to the pixel level proposing the following pixel value equation:

$$|pv_{(i,j,sl)}^{Acq\_\lambda}|^2 \cong A_{(i,j,sl)}^2 \exp\left(-\frac{TE_\lambda}{T_{2(i,j,sl)}^{(A)}/2}\right) + |N_{(i,j,sl)}|^2 \qquad \text{(Eq. E3.5–10)}$$

The validity of this approximation is further reinforced by noting that the signal-noise cross term decays twice as fast $(T_2/2)$ as the quadratic signal term, which decays with $T_2$.

Secondly, we review works on the effects of $B_0$ and $B_1$ inhomogeneities and of diffusion echo damping. Majumdar *et al.* studied the effects of $B_0$ inhomogeneities (Majumdar, Orphanoudakis, Gmitro, O'Donnell, & Gore, 1986b) as well as of $B_1$ imperfections (Majumdar, Orphanoudakis, Gmitro, O'Donnell, & Gore, 1986a) in $T_2$ qMRI with multi-SE pulse sequences. This group also studied the effects of intra-slice flip angle deviations caused by imperfect slice selective refocusing pulses (Majumdar & Gore, 1987). The echo damping effects of diffusion were studied (Deichmann *et al.*, 1995) and (Does & Gore, 2000). Altogether, the effects of these experimental imperfections can be described by extra attenuation factors for each echo of the train and the pixel values of the $\lambda^{th}$ echo are given by:

$$pv_{(i,j,sl)}^{Acq\_\lambda} = A_{(i,j,sl)} \exp\left(-\frac{TE_\lambda}{T_{2(i,j,sl)}^{(A)}}\right) (f_{(i,j,sl)}(\theta_{180}, \Delta B_0, T_2^{(A)}))^\lambda$$

$$\times \exp\left(-\frac{2}{3}\gamma^2 D_{(i,j,sl)} g_0^2 \lambda ES^3\right) \exp(i\phi_{\lambda(i,j,sl)}) \qquad \text{(Eq. E3.5–11)}$$

where $g_0$ is the background gradient (Deichmann, *et al.*, 1995; Does & Gore, 2000).

The extra attenuation factor f is a function of $\theta_{180}$ which is the deviation of the angle of rotation of the spins from a complete 180° and $\Delta B_0 = g_0 \Delta Z$ is the static field deviation from resonance. This perturbation factor of the echo amplitudes is an oscillatory function that is always less than one and depends on the specific envelope shape of the refocusing 180° pulses. The dependence of f on $T_2$ is not strong and reflects the manner in which longitudinal components artifactually produced by imperfect refocusing pulses later return to the transverse plane and decay between echoes (Majumdar & Gore, 1987).

The equation above implies that the problems of mapping $T_2$, D, and the experimental magnetic fields $B_0$ and $B_1$ become intertwined. As a first approximation, Sled and Pike (J. G. Sled & G. B. Pike, 2000) have studied the problem of $T_2$ quantification in the presence of $B_0$ and $B_1$ inhomogeneities *via*:

$$|pv_{(i,j,sl)}^{Acq-\lambda}| = A_{(i,j,sl)} \exp \left( -\frac{\lambda ES}{T_{2(i,j,sl)}^{(A)}} + \lambda \ln(f_{(i,j,sl)}) \right) \quad \text{(Eq. E3.5–12)}$$

where the diffusion effects have been neglected. In this case, an exponential fit to the measured pixel values $\{pv_{(i,j,sl)}^{Acq-\lambda}\}$ yields to an apparent or observed $T_2$ value, which in turn can be used to map the true $T_2$ with the equation:

$$T_{2(i,j,sl)}^{(A)} = \left( \frac{1}{T_{2,obs(i,j,sl)}^{(A)}} + \frac{\ln(f_{(i,j,sl)})}{ES} \right)^{-1} \quad \text{(Eq. E3.5–13)}$$

Such formalism has been used in conjunction with mappings of $B_0$ and $B_1$ generated in the same scanning session to correct the $T_2$ maps (J. G. Sled & G. B. Pike, 2000). To this end, these investigators make use of the formulas of the original theory (Majumdar, *et al.*, 1986a, 1986b) for the extra multiecho attenuation factor in terms of rotation matrices, to compensate for the effects of magnetic field inhomogeneities in $T_2$ qMRI.

### E3.5.ii. *Steady State Free Precession Techniques: qCV = FA*

As described in two recent papers (Deoni, *et al.*, 2003; Deoni, Ward, Peters, & Rutt, 2004), a different technique for $T_2$ qMRI uses images with mixed $T_2$- and $T_1$- weightings acquired with a steady state free precession pulse sequence, in conjunction with a spatially coregistered $T_1$ qMRI technique applied with the same geometric scanning settings. These investigators use the SPGR pulse sequence for $T_1$ (DESPOT1) and the SSFP pulse sequence for $T_2$. The theory supporting this combined technique stems from the

SSFP pixel value equation, specifically:

$$pv_{(i,j,sl)}^{Acq-\lambda} = \Gamma_{(i,j,sl)} PD_{(i,j,sl)}^{(A)}$$

$$\times \frac{\left(1 - \exp\left(-\frac{TR}{T_{1,(i,j,sl)}^{(A)}}\right)\right) \sin(FA^{Acq-\lambda}) \exp\left(-\frac{TE}{T_{2,(i,j,sl)}^{(A)*}}\right)}{\left(1 - \exp\left(-\frac{TR}{T_{1,(i,j,sl)}^{(A)}}\right) \exp\left(-\frac{TR}{T_{2,(i,j,sl)}^{(A)}}\right)\right.}$$

$$\left. - \cos(FA^{Acq-\lambda}) \left(\exp\left(-\frac{TR}{T_{1,(i,j,sl)}^{(A)}}\right) - \exp\left(-\frac{TR}{T_{2,(i,j,sl)}^{(A)}}\right)\right)\right)$$

$$(Eq.\ E3.5-14)$$

which can be rearranged into a form that is suitable for linear regression, specifically:

$$\frac{pv_{(i,j,sl)}^{Acq-\lambda}}{\sin(FA^{Acq-\lambda})} = \frac{pv_{(i,j,sl)}^{Acq-\lambda}}{tg(FA^{Acq-\lambda})} \frac{\left(\exp\left(-\frac{TR}{T_{1,(i,j,sl)}^{(A)}}\right) - \exp\left(-\frac{TR}{T_{2,(i,j,sl)}^{(A)}}\right)\right)}{\left(1 - \exp\left(-\frac{TR}{T_{1,(i,j,sl)}^{(A)}}\right) \exp\left(-\frac{TR}{T_{2,(i,j,sl)}^{(A)}}\right)\right)}$$

$$+ \Gamma_{(i,j,sl)} PD_{(i,j,sl)}^{(A)} \frac{\left(1 - \exp\left(-\frac{TR}{T_{1,(i,j,sl)}^{(A)}}\right)\right) \exp\left(-\frac{TE}{T_{2,(i,j,sl)}^{(A)*}}\right)}{\left(1 - \exp\left(-\frac{TR}{T_{1,(i,j,sl)}^{(A)}}\right) \exp\left(-\frac{TR}{T_{2,(i,j,sl)}^{(A)}}\right)\right)}$$

$$(Eq.\ E3.5-15)$$

which is of the general linear form:

$$Y_{(i,j,sl)} = m_{(i,j,sl)} X_{(i,j,sl)} + b_{(i,j,sl)} \qquad (Eq.\ E3.5-16)$$

where, as for DESPOT1 (see section E3.4iii), X and Y are parameterized as:

$$Y_{(i,j,sl)} = \frac{pv_{(i,j,sl)}^{Acq-\lambda}}{\sin(FA^{Acq-\lambda})} \qquad (Eq.\ E3.5-17)$$

and

$$X_{(i,j,sl)} = \frac{pv_{(i,j,sl)}^{Acq-\lambda}}{tg(FA^{Acq-\lambda})} \qquad (Eq.\ E3.5-18)$$

Consequently, linear regression allows calculating the slope, specifically:

$$m_{(i,j,sl)} = \frac{\left(\exp\left(-\frac{TR}{T_{1,(i,j,sl)}^{(A)}}\right) - \exp\left(-\frac{TR}{T_{2,(i,j,sl)}^{(A)}}\right)\right)}{\left(1 - \exp\left(-\frac{TR}{T_{1,(i,j,sl)}^{(A)}}\right)\exp\left(-\frac{TR}{T_{2,(i,j,sl)}^{(A)}}\right)\right)} \qquad \text{(Eq. E3.5–19)}$$

and the intercept which is given by:

$$b_{(i,j,sl)} = \Gamma_{(i,j,sl)}PD_{(i,j,sl)}^{(A)} \frac{\left(1 - \exp\left(-\frac{TR}{T_{1,(i,j,sl)}^{(A)}}\right)\right)\exp\left(-\frac{TE}{T_{2,(i,j,sl)}^{(A)*}}\right)}{\left(1 - \exp\left(-\frac{TR}{T_{1,(i,j,sl)}^{(A)}}\right)\exp\left(-\frac{TR}{T_{2,(i,j,sl)}^{(A)}}\right)\right)}$$

$$\text{(Eq. E3.5–20)}$$

In the final step, from the slope and the previously generated $T_1$ maps, one can map $T_2$ using the equation:

$$T_{2,(i,j,sl)}^{(A)} = -\frac{TR}{\ln\left(\frac{m_{(i,j,sl)} - \exp\left(-\frac{TR}{T_{1,(i,j,sl)}^{(A)}}\right)}{m_{(i,j,sl)}\exp\left(-\frac{TR}{T_{1,(i,j,sl)}^{(A)}}\right) - 1}\right)} \qquad \text{(Eq. E3.5–21)}$$

The SSFP pulse sequence is vulnerable to transverse magnetization interference artifacts that appear as dark bands in the images. These signal voids result from any condition that leads to the transverse magnetization incurring a net phase shift during the TR interval, including $B_0$ inhomogeneities and susceptibility differences, such as those seen near the sinuses and within the structures of the inner ear. A possible solution, which consist in cycling the phase of the rf pulses has been studied (Deoni, *et al.*, 2004) with encouraging results; the processing mathematics above remains as described above.

### E3.5.iii. *Fast Spin-echo Techniques: qCV = TE$_{eff}$*

Up to this point, we have considered $T_2$ qMRI techniques in which TE is a well-defined parameter for each directly acquired image. This is not the case when using hybrid-readout pulse sequences whereby different k-lines of an image are acquired at different sub-echo times, as for example in the case of RARE, FSE, and TSE pulse sequences described in section E2. With such pulse sequences, TE is no longer available as a qCV and is replaced with the effective echo time ($TE_{eff}$), which is the time at which the k-lines corresponding to phase encoding gradient amplitudes closest to zero, are acquired.

One important example is the dual-echo fast spin-echo (DE-FSE) pulse sequence (Melki, Mulkern, Panych, & Jolesz, 1991), which is widely used for research studies and increasingly in clinical practice. It generates two directly acquired images per slice, one acquired typically with a very short $TE1_{eff}$ and therefore with minimum $T_2$-weighting, and the second with a longer $TE2_{eff}$ typically chosen to produce a $T_2$-weighting close to that used clinically. The dual echo capability is achieved by dividing the FSE readout of even echo train length (ETL) into two sub echo trains of half the ETL each, and by using in general different phase encoding schemes --a.k.a. different profile orders-- for each. Typically, a centric profile order, in which the center k-lines are acquired at the beginning of the echo train are most suitable for achieving short $TE1_{eff}$ is used for the first echo. On the other hand, a linear profile order in which the center k-lines are acquired at the center of the train is more convenient for achieving the longer $TE2_{eff}$ needed for adequate $T_2$-weighting.

As discussed in the following, using different phase encoding profile orders for the first and second DE-FSE images has implications for mapping $T_2$ because the voxel sensitivity functions are not the same for the two echoes. The corresponding pixel value equations for the two echoes are:

$$\text{pv}^{\text{Acq\_1}}_{(i,j,sl)} = \left( \frac{\pi^2 \gamma^2 \hbar^2 B_0}{k_B T} \right) \Omega_{(i,j,sl)} \text{PD}^{(A)}_{(i,j,sl)} \text{psw}(\text{qMRI\_par}_{(i,j,sl)})$$

$$\times \exp \left( -\frac{TE1_{eff}}{T^{(A)}_{2,(i,j,sl)}} \right) \Delta V^{\text{Acq\_1}}_{(i,j,sl)} \qquad \text{(Eq. E3.5–22)}$$

and,

$$\text{pv}^{\text{Acq\_2}}_{(i,j,sl)} = \left( \frac{\pi^2 \gamma^2 \hbar^2 B_0}{k_B T} \right) \Omega_{(i,j,sl)} \text{PD}^{(A)}_{(i,j,sl)} \text{psw}(\text{qMRI\_par}_{(i,j,sl)})$$

$$\times \exp \left( -\frac{TE2_{eff}}{T^{(A)}_{2,(i,j,sl)}} \right) \Delta V^{\text{Acq\_2}}_{(i,j,sl)} \qquad \text{(Eq. E3.5–23)}$$

Therefore, $T_2$ can be mapped using the following equation:

$$T_{2,(i,j,sl)}^{(A)} = \frac{TE2_{eff} - TE1_{eff}}{\ln\left(\frac{pv_{(i,j,sl)}^{Acq\_1}\ \Delta V_{(i,j,sl)}^{Acq\_2}}{pv_{(i,j,sl)}^{Acq\_2}\ \Delta V_{(i,j,sl)}^{Acq\_1}}\right)} \qquad \text{(Eq. E3.5–24)}$$

This mapping equation is therefore very similar to the mapping equation of the conventional dual-echo spin-echo pulse sequences. The main differences are the voxel size ratio-factor in the denominator and the use of effective echo times in the numerator. In this equation, we have used the voxel volume equation:

$$\Delta V_{(i,j,sl)}^{Acq\_1,2} = \iiint\limits_{\left\{\begin{smallmatrix}\text{Infinite}\\\text{space}\end{smallmatrix}\right\}} VSF^{(Acq\_1,2)}(\vec{X}_{(i,j,sl)} - \vec{x})d^3x \qquad \text{(Eq. E3.5–25)}$$

as expressed in terms of the voxel sensitivity functions (see section C.5), which are given by:

$$VSF^{(Acq\_1,2)}(\vec{X}_{(i,j,sl)} - \vec{x}) \equiv \frac{S(X_{sl-x_{ss}})}{N_pN_f} \sum_{p,f=-N/2}^{N/2} \exp\left(\frac{TE1, 2_{eff} - te_{(f,p)}^{Acq\_1,2}}{T_2^{(A)}(\vec{X})}\right)$$

$$\times \exp[+2\pi i(\vec{X}_{(i,j)} - \vec{x}) \cdot \vec{K}_{(f,p)}] \qquad \text{(Eq. E3.5–26)}$$

Hence, the voxel volumes are not necessarily the same for both echoes because the sub-echo times are dependent on the profile order used. These effects are vanishingly small for long $T_2$ species in which case, the first exponential factor is near unity. In addition, these effects are negligible for conventional spin-echo pulse sequences assuming negligible $T_2$ decay during each frequency encoding signal readout. In both of these cases:

$$VSF^{(Acq\_1)}(\vec{X}_{(i,j,sl)} - \vec{x}) \cong VSF^{(Acq\_2)}(\vec{X}_{(i,j,sl)} - \vec{x}) \cong \frac{S(X_{sl-x_{ss}})}{N_pN_f} \sum_{p,f=-N/2}^{N/2}$$

$$\times \exp[+2\pi i(\vec{X}_{(i,j)} - \vec{x}) \cdot \vec{K}_{(f,p)}]$$

$$\text{(Eq. E3.5–27)}$$

and the mapping equation above reduces to the usual:

$$T_{2,(i,j,sl)}^{(A)} = \frac{TE2 - TE1}{\ln\left(\frac{pv_{(i,j,sl)}^{Acq\_1}}{pv_{(i,j,sl)}^{Acq\_2}}\right)} \qquad \text{(Eq. E3.5–28)}$$

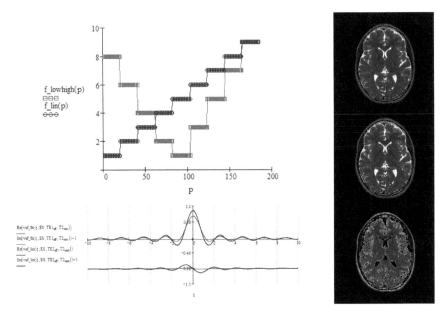

Fig. E-7.  $T_2$ mapping with dual echo TSE pulse sequences. Different TSE profile orders (top left) lead to different VSFs for the first echo (centric profile order) and the second (linear profile order). This has implications for $T_2$ mapping when using the full pixel value equations (see text). As shown in the uncorrected $T_2$ map (top right) and the corrected (middle right), the resulting $T_2$ maps are very similar visually but differ quantitatively: the difference map is shown at the bottom.

The profile order effect on $T_2$ qMRI is illustrated in Fig. E-7 for a centric (first echo) and a linear (second echo) profile orders. The voxel sensitivity function of the second echo is more sharply defined than that of the first echo, because there is a relative loss of spatial resolution with centric profile orders, with which the high frequency k-lines are read at later times than the center k-lines. In this particular example, the $T_2$ values obtained for an ROI in frontal white matter is about 10% shorter when using the voxel size correction-factor.

### E3.6.  *qMRI of the Reduced Transverse Relaxation Time $(T_2^*)$*

The standard method for generating $T_2^*$ maps uses a multiecho gradient-echo pulse sequence yielding gradient-echo images acquired at progressively long TEs. By recalling the gradient-echo pixel value equation adapted for describing multiple echoes, *i.e.*:

$$
\mathrm{pv}_{(i,j,sl)}^{Acq\text{-}\lambda} = \left( \frac{\pi^2 \gamma^2 \hbar^2 B_0}{k_B T} \right) \Omega_{(i,j,sl)} PD_{(i,j,sl)}^{(A)} |psw| \exp \left( -\frac{TE^{Acq\text{-}\lambda}}{T_{2,(i,j,sl)}^{(A)*}} \right)
$$

$$
\times \mathrm{sinc} \left( 2\pi\gamma \frac{\Delta B_0}{2} TE^{Acq\text{-}\lambda} \right) \exp(-i2\pi\gamma B_{0,(i,j,sl)} TE^{Acq\text{-}\lambda}) \Delta V_{(i,j,sl)}
$$

$$(\text{Eq. E3.6–1})$$

We notice that simple exponential decay is only a crude approximation and significant deviations can occur for regions with background gradients, particularly for longer TEs where the effects of the sinc function can become dominant. Methods for reducing the severity of these sources of error have been reported in the literature, as succinctly reviewed in a recent paper (Baudrexel *et al.*, 2009), these include: 1) increasing the spatial resolution of the scan (Young, Cox, Bryant, & Bydder, 1988) albeit at the expense of longer scan times and less SNR. 2) Using maps of the $B_0$ field for $T_2^*$ correction (An & Lin, 2002). 3) An iterative postprocessing approach (Fernández-Seara & Wehrli, 2000). 4) Methods that are based on the z-shimming technique, which recover signal amplitudes by applying appropriate counter gradient pulses, usually in the through-plane direction (Constable, 1995; Frahm, Merboldt, & Hänicke, 1988; Posse, Shen, Kiselev, & Kemna, 2003; Truong, Chakeres, Scharre, Beversdorf, & Schmalbrock, 2006; Wild, Martin, & Allen, 2002). And 5) a multi gradient-echo approach whereby the full complex images are used for direct calculation of suscepti-bility gradient maps rather than using this information for recursive fitting and these investigators use this information for correcting the modulus data (Baudrexel, *et al.*, 2009).

### E3.7. qMRI of the Semisolid Pool

In contrast to the liquid pool case, the qMRI parameters of the semisolid pool do not yet have known associations with individual pulse sequence qCVs on a parameter-by-parameter basis. Instead, the conventional tech-niques used for probing the semisolid pool derive from the magnetization transfer (MT) experiment (Wolff & Balaban, 1989) in which all the semisolid pool qMRI parameters affect to various degrees the pixel values of the generated MT-weighted images. MT weighting is achieved by driving the semisolid pool with off resonance rf radiation that is applied either in form of a prepulse or as a continuous wave (CW). In such experiments, the totality of the semisolid pool qMRI parameters become detectable as a group and not individually as is the case for liquid pool qMRI parameters. Accordingly, in the two-pool model (Edzes & Samulski, 1977; Grad, Mendelson, Hyder, & Bryant, 1991; Henkelman *et al.*, 1993), the parameters of the indirectly-detectable semisolid pool are the fundamental exchange constant R and,

$$
\text{qMRI\_par}^{(B)} = \left\{ \begin{array}{cccc} \text{PD}^{(B)} & \cdots & \cdots & \cdots \\ \vdots & \cdots & \cdots & \cdots \\ \vdots & \text{T}_1^{(B)} & \text{T}_2^{(B)} \end{array} \right\}
$$

$$\underbrace{\qquad\qquad\qquad\qquad\qquad\qquad\qquad}$$
$$\Uparrow$$
$$\text{qMRI Association}$$
$$\Downarrow$$
$$\overbrace{\text{MT\_prepulse or CW}}$$

(Eq. E3.7–1)

where we assumed that the spins of the semisolid pool have negligible kinetics, *i.e.* negligible diffusion, perfusion and flow. Another qMRI parameter, namely the semisolid pool fraction is frequently used instead of $\text{PD}^{(B)}$ or $\text{M}_z^{(B\_eq)}$ and is defined as follows:

$$
f \equiv \frac{\text{M}_z^{(B\_eq)}}{\text{M}_z^{(A\_eq)} + \text{M}_z^{(B\_eq)}} = \frac{\text{PD}^{(B)}}{\text{PD}^{(A)} + \text{PD}^{(B)}} \qquad \text{(Eq. E3.7–2)}
$$

where we have used the relationships:

$$
\text{PD}^{(A,B)} = \left( \frac{k_B T}{\pi^2 \gamma^2 \hbar^2 B_0} \right) \text{M}_z^{(A,B\_eq)} \qquad \text{(Eq. E3.7–3)}
$$

The magnetization transfer phenomenon can be made manifest on purpose by applying off-resonance rf radiation (Wolff & Balaban, 1989) -- either pulsed or continuous wave --, or it can manifest itself incidentally when using multislice (MS) pulse sequences (W. Dixon, Engels, Castillo, & Sardashti, 1990; Melki & Mulkern, 1992; Santyr, 1993). In this second case, the rf pulses used for exciting and refocusing the magnetizations of neighboring slices, unintentionally act as MT pulses; hence the more rf intensive the pulse sequence, the stronger the incidental MT effects (Melki & Mulkern, 1992; Weigel, Helms, & Hennig, 2010).

### E3.7.i. *MT-Bloch Equations*

Within the two-pool model, which is also known as the binary spin bath model, magnetization dynamics is governed by modified Bloch equations. For the transverse magnetizations of the two pools, these MT-Bloch equations (Graham & Henkelman, 1997; Henkelman, *et al.*, 1993) are:

$$\frac{\partial M_x^{(A,B)}}{\partial t} = -2\pi \Delta v M_y^{(A,B)} - \frac{M_x^{(A,B)}}{T_2^{(A,B)}} \qquad \text{(Eq. E3.7–4)}$$

and,

$$\frac{\partial M_y^{(A,B)}}{\partial t} = 2\pi \Delta v M_x^{(A,B)} - \frac{M_y^{(A,B)}}{T_2^{(A,B)}} - \omega_1(t) M_z^{(A,B)} \qquad \text{(Eq. E3.7–5)}$$

These two equations, which are fully symmetric for the two pools, do not explicitly include exchange terms for the transverse magnetization components. Such effects are assumed negligible due to the very short $T_2$ relaxation time of the semisolid pool. Furthermore, these equations can be combined to give the equations for the complex transverse magnetizations of either pool:

$$\frac{\partial m^{(A,B)}}{\partial t} = i2\pi \Delta v m^{(A,B)} - \frac{m^{(A,B)}}{T_2^{(A,B)}} - i\,\omega_1(t) M_z^{(A,B)} \qquad \text{(Eq. E3.7–6)}$$

In addition, the MT-Bloch equations for the longitudinal magnetizations of the two pools are:

$$\frac{\partial M_z^{(A)}}{\partial t} = \frac{(M_z^{(A\text{-}eq)} - M_z^{(A)})}{T_1^{(A)}} - RM_z^{(B\text{-}eq)}M_z^{(A)} + RM_z^{(A\text{-}eq)}M_z^{(B)}$$
$$+ \omega_1(t)M_y^{(A)} \qquad \text{(Eq. E3.7–7)}$$

and,

$$\frac{\partial M_z^{(B)}}{\partial t} = \frac{(M_z^{(B\text{-}eq)} - M_z^{(B)})}{T_1^{(B)}} - RM_z^{(A\text{-}eq)}M_z^{(B)} + RM_z^{(B\text{-}eq)}M_z^{(A)}$$
$$- R_{rfB}(\Delta v)M_z^{(B)} \qquad \text{(Eq. E3.7–8)}$$

In these equations, the time dependent amplitude of the $B_{1T}$ field has been expressed as an angular frequency *via* the standard relationship $\omega_1(t) = 2\pi\gamma B_{1T}(t)$. Furthermore, $R_{rfB}(\Delta v)$ is the rate of loss of longitudinal magnetization by the semisolid pool due to off-resonance irradiation of amplitude $\omega_1(t)$ and offset frequency $\Delta v$. We further recall that the pseudo-first-order rate constants for magnetization exchange processes in the forward and reverse directions are $RM_z^{(B\_eq)}$ : A $\longrightarrow$ B and $RM_z^{(A\_eq)}$ : B $\longrightarrow$ A, respectively.

The longitudinal magnetization differential equations above are symmetric to each other in all terms except for the rf excitation terms, since proper description of rf excitation of the semisolid pool requires using a non-Lorentzian lineshape (see section E3.7.iii below) (Henkelman, *et al.*, 1993; Morrison & Henkelman, 1995).

A modified set of MT-Bloch equations has been used by Sled and Pike (J. Sled & G. Pike, 2000; Sled & Pike, 2001). In this model, the mathematical description is considerably more involved because it leads to an additional differential equation, which is coupled to the MT-Bloch equations. Furthermore, this theoretical framework, which uses the Redfield-Provotorov theory and involves the concept of inverse spin temperature (J. Sled & G. Pike, 2000), leads to parametric map computations that involve solving the MT-Bloch equations numerically. Herein, this approach will not be discussed further; instead we will proceed along the lines of the work of Ramani *et al.* (Ramani, Dalton, Miller, Tofts, & Barker, 2002) and Tozer *et al.* (Tozer *et al.*, 2003). These investigators use the unmodified MT-Bloch equations above and extend the original steady state solutions of Henkelman *et al.* (Henkelman, *et al.*, 1993) to describe pulsed MT imaging experiments approximately. Recently, a comparative study was conducted (Portnoy & Stanisz, 2007) for the purpose of determining potential differences in qMT parameters as generated with the leading published qMT methodologies, specifically the methodologies of Ramani *et al.* (Ramani, *et al.*, 2002), Sled and Pike (Sled & Pike, 2001), and Yarnykh (Yarnykh, 2002). The results of this study demonstrated that the approximations used in pulsed MT modeling are quite robust. In particular, it was shown that the semisolid pool fraction $M_z^{(B\_eq)}$ and $T_2^{(B)}$ could be evaluated with reasonable accuracy regardless of the model used (Portnoy & Stanisz, 2007).

E3.7.ii. *The Steady State*

In the steady state, all time derivatives are null, accordingly:

$$\frac{\partial M_x^{(A,B)}}{\partial t} = \frac{\partial M_y^{(A,B)}}{\partial t} = \frac{\partial M_z^{(A,B)}}{\partial t} = 0 \qquad \text{(Eq. E3.7–9)}$$

Therefore, the MT-Bloch equations lead to:

$$0 = -\frac{M_x^{(A,B)}}{T_2^{(A,B)}} - 2\pi\Delta v M_y^{(A,B)} \tag{Eq. E3.7-10}$$

and,

$$0 = -\frac{M_y^{(A,B)}}{T_2^{(A,B)}} + 2\pi\Delta v M_x^{(A,B)} - \omega_1(t)M_z^{(A,B)} \tag{Eq. E3.7-11}$$

These two equations can be solved for $M_x^{(A,B)}$ and $M_y^{(A,B)}$, therefore leading to (Henkelman, *et al.*, 1993):

$$0 = \frac{(M_z^{(A\_eq)} - M_z^{(A)})}{T_1^{(A)}} - RM_z^{(B\_eq)}M_z^{(A)} + RM_z^{(A\_eq)}M_z^{(B)}$$

$$- \frac{\omega_1^2 T_2^{(A)}}{1 + 2\pi\Delta v T_2^{(A)}}M_z^{(A)} \tag{Eq. E3.7-12}$$

and,

$$0 = \frac{(M_z^{(B\_eq)} - M_z^{(B)})}{T_1^{(B)}} - RM_z^{(A\_eq)}M_z^{(B)} + RM_z^{(B\_eq)}M_z^{(A)}$$

$$- R_{rfB}(\Delta v)M_z^{(A\_eq)}M_z^{(B)} \tag{Eq. E3.7-13}$$

In turn, these coupled algebraic equations can be solved for $M_z^{(A)}$. With this solution, a pixel value equation can be derived for MT imaging experiments whereby several images are acquired with different frequency offsets $\Delta v^{Acq\text{-}\alpha}$. Adapting the published result (Henkelman, *et al.*, 1993; Ramani, *et al.*, 2002) to the notation used in this book and using the fact that $(2\pi\Delta v T_2^{(A)})^2 << 1$, the pixel value equation is:

$$pv^{Acq\text{-}\alpha}$$

$$= \left( \frac{\kappa M_z^{(A\_eq)}\left(\frac{1}{T_1^{(B)}}\left(RM_z^{(A\_eq)}T_1^{(A)}\frac{f}{(1-f)}\right)\right.}{\left(\frac{RM_z^{(A\_eq)}T_1^{(A)}f}{(1-f)}\right)\left(\frac{1}{T_1^{(B)}} + R_{rfB}(\Delta v^{Acq\text{-}\alpha})\right) + \left(1 + \left(\frac{\omega_{1CWPE}}{2\pi\Delta v_\alpha}\right)^2\right)} \right.$$

$$\left. \frac{\left. + R_{rfB}(\Delta v^{Acq\text{-}\alpha}) + \frac{1}{T_1^{(B)}} + RM_z^{(A\_eq)}\right)}{\times \left(\frac{T_2^{(A)}}{T_1^{(B)}}\right)\right)\left(R_{rfB}(\Delta v^{Acq\text{-}\alpha}) + \frac{1}{T_1^{(B)}} + RM_z^{(A\_eq)}\right)} \right)$$

$$\tag{Eq. E3.7-14}$$

where the continuous wave power equivalent is given by:

$$\omega_{1CWPE} = 2\pi\gamma B_{1CWPE} \qquad \text{(Eq. E3.7–15)}$$

The concept of continuous wave power equivalent was introduced for the purpose of adapting this steady state model to the pulsed case (Ramani, *et al.*, 2002). Although the pixel value equation above is strictly valid only under steady state conditions, it is nevertheless used for analyzing inherently dynamic data, specifically for images acquired with MT-prepulses.

### E3.7.iii. *The Bound Water Lineshape*

The rf absorption rate of the semisolid pool can be expressed in terms of a generalized absorption lineshape function $g(2\pi\Delta\nu T_2^{(B)})$ *via*:

$$R_{rfB} = 4\pi^2\gamma^2 B_1^2 g(2\pi\Delta\nu T_2^{(B)}) \qquad \text{(Eq. E3.7–16)}$$

which is a quadratic function of the applied off resonance $B_1$ field. The corresponding expression for the pulsed case, as is of interest to clinical applications, has been modeled in terms of a "continuous wave power equivalent" magnetic field and is given by (Ramani, *et al.*, 2002):

$$B_1 \xrightarrow{\text{pulsed MT}} B_{1CWPE} = B_{SAT}\sqrt{p\frac{\tau_{SAT}}{TR'}} \qquad \text{(Eq. E3.7–17)}$$

In this equation $B_{SAT}$ and $\tau_{SAT}$ are the amplitude and duration of the MT pulse. For a 3DFT sequence, or a single-slice 2DFT sequence, TR' is equal to the sequence repetition time, TR. However, in the case of multislice 2DFT sequences, TR' is equal to the repetition time divided by the number of slices collected per TR. Finally, p is the ratio of the square of the mean amplitude of the saturation pulse to that of a rectangular pulse of the same height (Ramani, *et al.*, 2002).

The Lorentzian lineshape is associated with spin packets where the spin-spin dipolar interactions are averaged out due to motions (*i.e.*, liquids), whereas the Gaussian lineshape is associated with spins in a less mobile system where the spin-spin dipolar interactions are approximately static. Accordingly, the rates of rf absorption rates by the liquid and semisolid pools can be modeled respectively as follows:

$$R_{rfA} = 4\pi^3\gamma^2 B_{1CWPE}^2 \left\{ \frac{T_2^{(A)}}{\pi} \frac{1}{1 + (2\pi\Delta\nu T_2^{(A)})^2} \right\} \qquad \text{(Eq. E3.7–18)}$$

and,

$$R_{rfB}(\Delta\nu) = 4\pi^3\gamma^2 B_{1CWPE}^2 \left\{ \frac{T_2^{(B)}}{\sqrt{2\pi}} \exp\left( -\frac{(2\pi\Delta\nu T_2^{(B)})^2}{2} \right) \right\}$$

(Eq. E3.7–19)

Alternatively, a superLorentzian lineshape of the form:

$$R_{rfB}(\Delta\nu) = 4\pi^3\gamma^2 B_{1CWPE}^2 \left\{ \sqrt{\frac{2}{\pi}} \int_0^{\pi/2} \sin(\theta) \frac{T_2^{(B)}}{|3\cos^2(\theta) - 1|} \right.$$

$$\left. \times \exp\left( -2 \left( \frac{2\pi\Delta\nu T_2^{(B)}}{|3\cos^2(\theta) - 1|} \right)^2 \right) d\theta \right\} \qquad \text{(Eq. E3.7–20)}$$

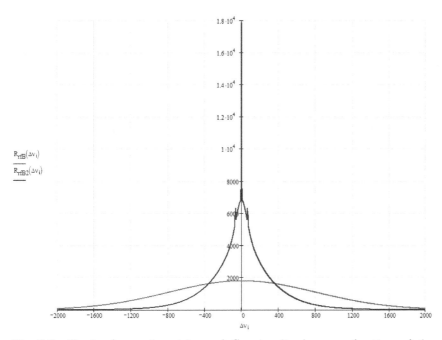

Fig. E-8. Graph of super-Lorentzian and Gaussian lineshapes as functions of the frequency offset. Parameters used for this simulations: $T_2^{(A)} = 100\,\text{ms}$; $T_2^{(B)} = 200\,\mu\text{s}$. Note that the expression for the super-Lorentzian lineshape has a singularity at the Larmor resonance frequency $\Delta\nu = 0$. It is believed however that other interactions not represented in this model limit its height to an extent that the effect on the restricted pool of the small angle on-resonance pulses used in the imaging experiments is negligible (Sled & Pike, 2001).

has been found to more accurately describe the semisolid pool of several tissues (Morrison, Stanisz, & Henkelman, 1995). As pointed out by Morrison and Henkelman, the super-Lorentzian is the lineshape that is expected to arise from partially ordered materials, for example, polymers, liquid crystals, biological membranes or molecules absorbed on surfaces (Morrison & Henkelman, 1995). The super-Lorentzian lineshape was first observed in potassium laurate and subsequently in lamellar liquid crystals and lipids. It has also been seen as part of the complex proton NMR spectrum of whole cells. Understanding the physical origin of this lineshape requires theoretical developments that are beyond the scope of this book; the reader is referred to the paper by Morrison, Stanisz and Henkelman (Morrison, *et al.*, 1995). Other lineshapes based on the actual fits to the data have also been studied (Li, Graham, & Henkelman, 1997).

### E3.7.iv. *Model Parameters*

Altogether, this (semi) steady state MT pixel value model has six parameters, specifically: R, $M_z^{(A-eq)}$, f, $T_1^{(A)}, T_1^{(B)} T_2^{(B)}$. Four of these parameters can be determined by fitting the functions of Eq. E3.7-13 to the experimental pixel values of several MT-weighted images obtained with different frequency offsets; the exceptions being the equilibrium longitudinal magnetization of the liquid pool, which is set to $M_z^{(A-eq)} = 1$ and $T_1^{(A)}$ which is measured with a separate $T_1$-qMRI experiment without off-resonance irradiation. This provides a measurement of the observed longitudinal relaxation time $T_1^{(A,obs)}$ of the liquid pool, which is related to the parameters of the semisolid pool *via*:

$$\frac{1}{T_1^{(A)}} = \frac{\frac{1}{T_1^{(A,obs)}}}{1 + \left(\frac{RM_z^{(B-eq)}T_1^{(A)}\left(\frac{1}{T_1^{(B)}} - \frac{1}{T_1^{(A,obs)}}\right)}{\frac{1}{T_1^{(B)}} - \frac{1}{T_1^{(A,obs)}} + R}\right)} \qquad \text{(Eq. E3.7–21)}$$

This equation can be derived by solving the MT-Bloch equations in the transient regime and without off-resonance radiation (Henkelman, *et al.*, 1993).

To gain insight into the mathematical characteristics of the binary spin bath model, we study by numerical simulation the effects of individually varying the various model parameters –*i.e.* R, f, $T_1^{(A)}$, $T_1^{(B)}$, $T_2^{(B)}$– while keeping all other parameter values fixed. To this end, we evaluate

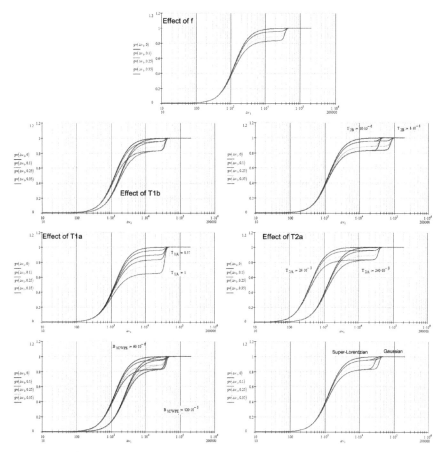

Fig. E-9. Effects of individually varying the various model parameters −*i.e.* R, f, $T_1^{(A)}, T_1^{(B)}, T_2^{(B)}$− while keeping all other parameter values fixed (see text).

Eq. E3.7-13 using a gaussian lineshape for the semisolid pool. First, we study the effect of the increasing the semisolid fraction (Fig. E-10), starting from a value of zero, which represents a material that does not exhibit MT effects, such as pure water, CSF, olive oil, and adipose: for f = 0, the pixel value as function of frequency offset curves resemble a tilted integral sign with flat ends. Then, for substances that do exhibit MT effects (f > 0), the curves develop a distinctive shoulder towards lower values starting at high offset frequencies. Most notably, only the depth of the shoulder increases as function of increasing f while the starting point of the shoulder remains constant as function of f. A similar behavior is obtained by varying the fundamental MT constant R. On the other hand, the position of the shoulder's starting point is strongly dependent on the value of $T_2^{(B)}$ as illustrated in (Fig. E-10); the starting point of shifts to lower offset frequencies as a function of increasing $T_2^{(B)}$. Interestingly, the starting point of the shoulder also depends on the lineshape of the semisolid pool (see Fig. E-10).

### E3.7.v. *Incidental MT Effects in 2D Multislice Imaging*

As was mentioned earlier, MT effects are also observed when using 2D-multislice pulse sequences. This unintended manifestation of the MT phenomenon is increasing attenuation of tissue signals as function of increasing number of slices (W. Dixon, *et al.*, 1990), and these effects are more prominent for rf intensive pulse sequences such as fast spin-echo (Melki & Mulkern, 1992), which play out several refocusing pulses during signal readout. Hence, the incidental MT effect with FSE pulse sequences increases with the echo-train-length (ETL) of the readout as well as with the number of slices of the scan. As can be expected, the incidental MT effect decreases for FSE readouts with reduced flip angles (Weigel, *et al.*, 2010).

The multislice attenuation effect has been modeled as a function of the number of slices (Melki & Mulkern, 1992) for the purpose of extracting the forward pseudo-first-order rate constant $RM_z^{(B\_eq)}$. To this end, the attenuation equation (Eng, Ceckler, & Balaban, 1991):

$$\frac{pv^{MS}}{pv^{ss}} = A + \frac{1 - A}{1 + \alpha B_1^2} \qquad \text{(Eq. E3.7–22)}$$

where MS and SS stand for the multislice and single slice cases respectively, and

$$\alpha = \frac{2\pi^2 \gamma^2 T_1^{(B)} T_2^{(B)}}{1 + \Delta v^2 T_2^{(B)2}} \qquad \text{(Eq. E3.7–23)}$$

and,

$$A = \frac{1}{1 + RM_z^{(B\_eq)} T_1^{(A)}} \qquad \text{(Eq. E3.7–24)}$$

was modeled as a function of the number of slices (N):

$$\frac{pv^{MS}}{pv^{SS}} = A + \frac{1 - A}{1 + \alpha'(N - 1)} \qquad \text{(Eq. E3.7–25)}$$

and obtained excellent agreement with brain data obtained as a function of increasing number of slices thus reinforcing the notion that rf pulses for adjacent slices in fact act as MT prepulses.

## E3.8.  qMRI of Fat

### E3.8.i.  *Introduction*

The preceding sections of this chapter where written with $^1$H-protons of aqueous tissue in mind. However, many of the principles formulated above can be extended to voxels containing fat only and, as was done for aqueous tissues, we can represent fat by its own set of qMRI parameters:

$$\text{qMRI\_par}^{(F)} = \begin{cases} \text{PD}^{(F)} & \cdots & \cdots & \cdots \\ \vdots & \text{D}^{(F)} & (\text{f}^{(F)}, \text{D}^{*(F)}) & \vec{V}^{(F)} \\ \vdots & \text{T}_1^{(F)} & (\text{T}_2^{(F)}, \text{T}_2^{*(F)}) \end{cases} \qquad \text{(Eq. E3.8–1)}$$

In some respects, qMRI of fat --for mapping the parameters above-- appears comparatively simpler than qMRI of aqueous tissues because fat has a very small diffusion coefficient of about two orders of magnitude slower than diffusion water (Ababneh *et al.*, 2009; Lehnert, Machann, Helms, Claussen, & Schick, 2004). Additionally, fat exhibits very weak MT effects (Weigel, *et al.*, 2010); it appears that adipose tissue, fat, and oils used in phantoms, behave as single spin pool systems and therefore magnetization exchange phenomena may not play a significant role in current MRI techniques. For the sake of completeness and symmetry, the perfusion qMRI parameters where kept in the expression above, although these effects may be negligibly small for fat. Hence, from this simplified perspective, clinical qMRI of fat involves the quantification of fewer parameters:

$$\text{qMRI\_par}^{(F)} \approx \begin{cases} \text{PD}^{(F)} & \cdots & \cdots & \cdots \\ \vdots & & & \\ \vdots & \text{T}_1^{(F)} & (\text{T}_2^{(F)}, \text{T}_2^{*(F)}) \end{cases} \qquad \text{(Eq. E3.8–2)}$$

On the other hand, in other respects qMRI of fat can be considerably more intricate than qMRI of aqueous tissue because fat has a multi-peak $^1$H-proton spectrum. Accordingly, there are in principle, as many sets of qMRI parameters –*i.e.* Eq. E3.8-2– as there are peaks in the spectrum of adipose.

Human adipose tissue is composed largely of triglycerides. Triglycerides consist of a glycerol molecule $(C_3H_5(OH)_3)$ bound to three fatty acids: seven fatty acids predominate and account for well over 90% of the fatty acids in human adipose tissue (Ren, Dimitrov, Sherry, & Malloy, 2008). These are myristic (3%), palmitic (19–24%), palmitoleic (6–7%), stearic (3–6%), oleic (45–50%), linoleic (1–2%) and linolenic (1-2%). The chemical formula of a generic triglyceride (Ren, *et al.*, 2008) is:

$$CH_2 - O - C(O) - R$$

$$|$$

$$CH - O - C(O) - R' \qquad \text{(Eq. E3.8–3)}$$

$$|$$

$$CH_2 - O - C(O) - R''$$

where R, R', and R'' designate fatty acids.

Ten resonances have been resolved for subcutaneous fat and fatty bone marrow of the tibia at 7T (Ren, *et al.*, 2008). The largest spectral peak corresponds to the methylene groups $(CH_2)_n$ and is shifted downfield by approximately 3.5 ppm of the water peak. Nine other resonances include $^1$H-protons of different chemical groups along the fatty acid. Fatty-hematopoietic bone marrow have similar spectra and have a prominent water resonance (Hilaire, Wehrli, & Song, 2000).

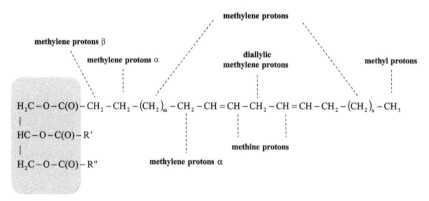

Fig. E-10.   The chemical structure of triglycerides: glycerol molecule $(C_3H_5(OH)_3)$ bound to three fatty acids.

The existence of multiple resonances can lead to signal interference phenomena, which is characteristic of systems with multiple natural frequencies. In particular, the MR signal interference effects caused by the 3.5 ppm spectral difference between the water and the methylene resonances is useful for designing fat-water separation imaging techniques as well as fat quantification techniques. Fat-water separation techniques provide separate images of either the fat or the water distributions inside the patient. However, besides being spectrally specific to either fat or water, the generated images are not necessarily quantitative in the qMRI sense because the pixel values are in general qualitative. More specifically, such pixel values are relative numbers that are proportional to the fat or water proton densities weighted by some combination of relaxation times, and these depend on the pulse sequence used. In a strict qMRI sense, fat quantification means generating maps of the proton density of fat, ideally including all spectral resonances. In practice, quantitative imaging of fat can be achieved by modeling the fat-water interference effects of the signal decay curves as a function of time (TE). Such chemical shift or Dixon type techniques (W. T. Dixon, 1984) have evolved continuously leading techniques that are very robust to $B_0$ magnetic field inhomogeneities (Glover, 1991; Glover & Schneider, 1991; Skinner & Glover, 1997). Furthermore, these have recently received much attention with the development of the iterative decomposition of water and fat with echo asymmetry and least square estimation (IDEAL) (Reeder et al., 2005; Reeder et al., 2004). This iterative reconstruction technique was initially developed with fat being modeled as a single resonance (Reeder, Hargreaves, Yu, & Brittain, 2005; Reeder, Pineda, et al., 2005; Reeder, et al., 2004) --the principal methylene resonance-- and later was extended to account for the multi-peak spectrum of fat (Yu et al., 2008). The IDEAL fat quantification algorithm can be used to process images acquired at multiple asymmetric TEs and with either of several chemical shift pulse sequences, including gradient-echo, fast spin-echo, gradient and spin-echo (GRASE); here we focus on the spoiled gradient-echo (SPGR) pulse sequence (Liu, McKenzie, Yu, Brittain, & Reeder, 2007). Most of the discussion however applies to all pulse sequences adapted for IDEAL reconstruction, as referenced above. This technique combines the acquisition of echoes acquired asymmetrically and an iterative least-squares water–fat separation decomposition algorithm in order to maximize the noise performance of the water–fat decomposition (Reeder, Pineda, et al., 2005). In one implementation of this technique (Liu, et al., 2007), the generated SPGR-IDEAL data sets covered 15 TE values with a spacing of

1.2 ms. The described IDEAL reconstruction algorithm had provisions for correcting the potentially confounding factors of: 1) $T_1$-bias whereby the different relaxation effects of water and fat were either taken into account mathematically, or minimized by using an SPGR sequence with very small flip angle such that $\cos(FA) \approx 1$ (typically FA = 5°).2) Used a magnitude discrimination method to calculate fat fraction without noise bias, and 3) included corrections for overall $T_2^*$ decay. In a more recent implementation, provisions for correcting independently for the $T_2^*$ decay of water and fat has been described (Chebrolu *et al.*, 2010).

### E3.8.ii. *Fat-Water Interference Phenomena*

In the developments that follow, we will model fat by the methylene resonance only and make use of the fact that these spins precess at the water Larmor frequency minus the chemical shift:

$$\Delta\omega_{(CS)} = 2\pi\gamma B_0 \Delta\delta[\text{ppm}]10^{-6} \qquad \text{(Eq. E3.8–4)}$$

Such angular frequency shift is magnetic field and temperature dependent; for a 1.5 T scanner and at body temperature, it is approximately 1,318 radian-Hz (or 210 Hz).

Water-fat mixture voxels are present at the interfaces between aqueous organs and body fat due to partial voluming, and can be found in organs affected by certain disease conditions such as nonalcoholic fatty liver disease (NAFLD) in which the liver becomes fatty infiltrated to varying degrees depending on disease severity. The main difference between voxels with partial volume *versus* fatty infiltrated voxels relates to the intravoxel spatial distribution of fat. In the former, fat occupies a subvoxel compartment and in the latter, fat and water are homogeneously distributed and both tissues occupy the entire voxel.

Theoretical qMRI analysis of voxels containing both water and fat $^1$H-protons either by partial volume or in diffuse water-fat mixtures is still an area of active investigation. The fundamental question relates to understanding how the individual subvoxel-qNMR parameters of water and fat combine mathematically to yield the voxel qMRI parameter. One can logically assume that the individual proton densities combine algebraically, that is:

$$PD_{(i,j,sl)} = PD_{(i,j,sl)}^{(A)} + PD_{(i,j,sl)}^{(F)} \qquad \text{(Eq. E3.8–5)}$$

where, for consistency of notation with other sections of this book, we continue using the superscript (A) to designate the liquid (water) pool of aqueous tissue. We further define the fat fraction as:

$$f_{(i,j,sl)} = \frac{PD^{(F)}_{(i,j,sl)}}{PD^{(A)}_{(i,j,sl)} + PD^{(F)}_{(i,j,sl)}} = \frac{PD^{(F)}_{(i,j,sl)}}{PD_{(i,j,sl)}} \qquad \text{(Eq. E3.8–6)}$$

and the proton densities are related by:

$$PD^{(F)}_{(i,j,sl)} = PD^{(A)}_{(i,j,sl)} \left( \frac{f_{(i,j,sl)}}{1 - f_{(i,j,sl)}} \right) \qquad \text{(Eq. E3.8–7)}$$

To illustrate methods of water-fat qMRI decomposition and quantification, we will analyze the problem of diffuse fatty infiltration in which both water and fat $^1$H-protons are distributed homogeneously inside the voxel; in this case, both water and fat occupy the entire voxel and the pixel value equation for a gradient-echo acquisition reads:

$$pv_{(i,j,sl)} = \frac{\Gamma_{(i,j,sl)}}{2} \left\{ \begin{array}{l} PD^{(A)}_{(i,j,sl)} psw(qCV \| qMRI\_par^{(A)}_{(i,j,sl)}) \exp\left( -\frac{TE}{T^{*(A)}_{2,(i,j,sl)}} \right) \\[2ex] + PD^{(F)}_{(i,j,sl)} psw(qCV \| qMRI\_par^{(F)}_{(i,j,sl)}) \\[2ex] \times \exp(i\Delta\omega_{(CS)}TE) \exp\left( -\frac{TE}{T^{*(F)}_{2,(i,j,sl)}} \right) \end{array} \right\}$$

$$\text{(Eq. E3.8–8)}$$

where we have assumed that the voxel contains an aqueous tissue that is diffusely fatty interspersed by a certain fat fraction (f) and the factor of one-half is necessary in order not to count the voxel volume twice. Furthermore, for gradient-echo acquisitions, the pulse sequence weighting factors for water and fat are functions only of $qCV = TR$ and $qMRI\_par = T_1$ and are given by:

$$psw(qCV \| qMRI\_par^{(A,F)}_{(i,j,sl)}$$

$$= \exp(-i2\pi\gamma\delta B_{0,(i,j,sl)}TE) \sin(FA) \left( \frac{1 - \exp\left( -\frac{TR}{T^{(A,F)}_{1,(i,j,sl)}} \right)}{1 - \cos(FA) \exp\left( -\frac{TR}{T^{(A,F)}_{1,(i,j,sl)}} \right)} \right)$$

$$\text{(Eq. E3.8–9)}$$

Hence, the resulting pixel value equation for water-fat mixture voxels is:

$$pv_{i,j,sl} = \frac{\Gamma_{(i,j,sl)}}{2} \sin(FA) \exp(-i2\pi\gamma\delta B_{0,(i,j,sl)}TE)PD^{(A)}_{(i,j,sl)}$$

$$\times \left\{ \begin{array}{l} \left( \dfrac{1-\exp\left(-\frac{TR}{T^{(A)}_{1,(i,j,sl)}}\right)}{1-\cos(FA)\exp\left(-\frac{TR}{T^{(A)}_{1,(i,j,sl)}}\right)} \right) \exp\left(-\frac{TE}{T^{*(A)}_{2,(i,j,sl)}}\right) \\[2em] + \left( \dfrac{f_{(i,j,sl)}}{1-f_{(i,j,sl)}} \right) \dfrac{1-\exp\left(-\frac{TR}{T^{(F)}_{1,(i,j,sl)}}\right)}{1-\cos(FA)\exp\left(-\frac{TR}{T^{(F)}_{1,(i,j,sl)}}\right)} \\[2em] \times \exp\left(i\Delta\omega_{(CS)}TE\right)\exp\left(-\frac{TE}{T^{*(F)}_{2,(i,j,sl)}}\right) \end{array} \right\}$$

(Eq. E3.8–10)

When studied as a function of TE, the modulus of the water-fat pixel value equation above gives rise to a function with a series of interleaved maxima and minima, and these occur at echo times that are multiple of chemical shift half period, *i.e.*:

$$\frac{\pi}{\Delta\omega_{(CS)}} = \begin{cases} 2.38\,\text{ms} & \text{at 1.5T} \\ 1.19\,\text{ms} & \text{at 3.0T} \end{cases}$$

(Eq. E3.8–11)

accordingly, the minima and maxima occur at:

$$TE_n = n\frac{\pi}{\Delta\omega_{(CS)}} \longleftrightarrow \begin{cases} \text{destructive interference(odd } n = 1, 3, \ldots) \\ \text{constructive interference(even } n = 2, 4, \ldots) \end{cases}$$

(Eq. E3.8–12)

As shown by computer simulation in Fig. E-12 below, the value of the fat fraction parameter has a marked effect on the pixel value minima and almost no effect on the maxima. This phenomenon of fat fraction dependent partial destructive interference is therefore useful for determining the fat fraction with qMRI techniques whereby several images are acquired at different TEs and these are referred generically as chemical shift fat quantification techniques. The common basic idea of these is to sample judiciously the pixel value curve so to have enough information to fit best a theoretical pixel value curve, thus leading to a value of the fat fraction. In addition to the fat fraction $f_{(i,j,sl)}$, the gradient-echo pixel value equation above has seven other unknowns per pixel, specifically the main qMRI parameters of the isolated water and fat components $PD^{(A,F)}_{(i,j,sl)}$, $T^{(A,F)}_{1,(i,j,sl)}$,

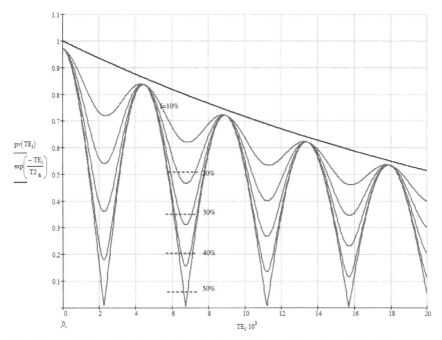

$$\frac{\mathrm{pv}(\mathrm{TE_i})}{\exp\left(\dfrac{-\mathrm{TE_i}}{\mathrm{T2_a}}\right)}$$

$\mathrm{TE_i} \cdot 10^3$

Fig. E-11. Effect of fat fraction in the pixel values as function of TE. Five curves are compared with fat fractions ranging from 10%, 20%, 30%, 40%, and 50%. All curves exhibit the same periodicity but the magnitude of the destructive interference as expected, increases with fat fraction up to 50% where destructive interference is maximum, *i.e.* null pv.

$T_{2,(i,j,sl)}^{*(A,F)}$ and the inhomogeneity field $\delta B_{0,(i,j,sl)}$ and, all of these should be known *a priory* or determined experimentally for accurate quantification of the fat fraction parameter.

### E3.9. *qMRI of Temperature*

Most qMRI parameters are temperature dependent and several have been used as proxies for indirect temperature mapping, as discussed in a recent review article on MR thermometry (Rieke & Pauly, 2008b). The primary clinical application of MR thermometry is monitoring minimally invasive thermal therapy of benign and malignant disease with the goal of using real-time temperature mapping to provide more control over the treatment outcome. Although there are in principle as many MRI thermometry techniques as temperature dependent qMRI parameters, the technique that best meets the clinical requirements of speed (real time) and temperature sensitivity is based on the temperature dependence of the $^1$H-proton resonance frequency (PRF). This is a magnetic shielding phenomenon originally observed in the context of studying hydrogen bond formation between water molecules (Hindman, 1966). It was later adapted for MRI by two separate groups using a spin-echo (Ishihara *et al.*, 1995) and gradient-echo pulse sequences (De Poorter, 1995; De Poorter *et al.*, 1995) respectively.

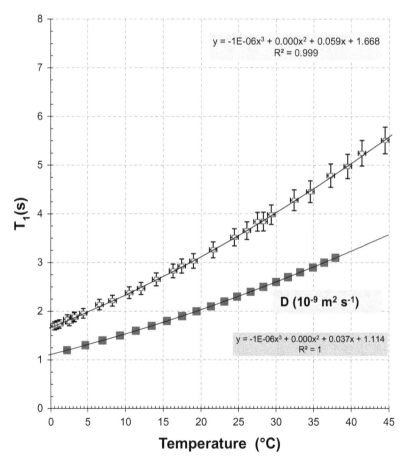

Fig. E-12. Temperature dependencies of $T_1$ and diffusion coefficient of pure deoxygenated water. Graphed using published experimental data in (Hindman, Svirmickas, & Wood, 1973; P. S. Tofts *et al.*, 2008).

E3.9.i. *Proton Resonance Frequency (PRF)*

The $^1$H protons of an isolated water molecule are more shielded by the electronic cloud that the $^1$H-protons of liquid water and aqueous tissue. In the liquid phase, hydrogen bonds develop between neighboring molecules and these distort the electronic configuration of the individual molecules resulting in a reduction of the electronic screening; hence leading to higher PRF. Because the fraction and the nature of the hydrogen bonds in water vary with temperature (Némethy & Scheraga, 1962), there is a temperature dependent mean PRF. As the temperature of the liquid water increases, the hydrogen bonds stretch, bend (Hindman, 1966), and break (Schneider, Bernstein, & Pople, 1958), and consequently on average the water molecules spend less time in a hydrogen-bonded state that is less shielded. Macroscopically, this can be detected as a temperature dependent phase shift, which is measurable with suitable pulse sequences and processing algorithms.

The temperature dependent contribution to the local microscopic magnetic field at position $\vec{x}$ is approximately given by:

$$\delta \vec{B}^{(loc)}(\vec{x}, t) \cong - \left( \frac{2}{3} \chi(\vec{x}, T) + \sigma(\vec{x}, T) \right) B_0 \, \hat{z} \quad \text{(Eq. E3.9–1)}$$

where higher order contributions have been neglected, and $\chi$ and $\sigma$ are the volume susceptibility of tissue and the shielding constant of the protons, respectively. These two quantities are temperature dependent and for the temperature range of interest, *e.g.* 30°C to 45°C, both of these can be approximated as linear functions of the temperature. The main difference between these however is that the proportionality constant is much higher for the shielding constant than for the volume susceptibility, and in addition, the shielding proportionality constant is largely tissue independent for aqueous tissue and water, the exception being adipose (Rieke & Pauly, 2008a). For this reasons, most implementations of PRF thermometry in aqueous tissue assume only temperature effects of the shielding constant (Rieke & Pauly, 2008b).

$$\sigma(\vec{x}_{(i,j,sl)}, T_{(i,j,sl)}) = \alpha T_{(i,j,sl)} \quad \text{(Eq. E3.9–2)}$$

where $\alpha$ being tissue independent, is therefore assumed to be position independent. Hence, if the phase maps of two acquisitions performed at different temperatures are subtracted, the temperature-difference can be mapped with the following equation:

$$\Delta T_{(i,j,sl)} = \frac{\phi_{(i,j,sl)}(T) - \phi_{(i,j,sl)}(T_0)}{2\pi\gamma\alpha B_0\, TE} \qquad \text{(Eq. E3.9–3)}$$

In practice (Peters, Hinks, & Henkelman, 1998), at 1.5 T, a PRF-thermal coefficient of $-0.01$ ppm/°C produces a phase shift of 4.6 phase degrees/°C over a TE of 20 ms.

### E3.9.ii. *Temperature Mapping via Proton Density*

The equilibrium longitudinal magnetization and the proton density exhibit distinct temperature dependences:

$$M_z^{(eq)}(\vec{x}, T) = \left(\frac{\pi^2\gamma^2\hbar^2 B_0}{k_B T}\right) rPD(\vec{x})\frac{NP}{MW}\rho_w(T) \qquad \text{(Eq. E3.9–4)}$$

where, as discussed in section E3.1, NP and MW are the number of protons and the molecular weight of the water molecule and rPD is the water relative proton density. Furthermore, the physical density of liquid water is to a very good approximation given by:

$$\rho_w(T) \cong 10^3(3.651\ 10^{-8}T^3 - 7.439\ 10^{-6}T^2 + 4.862\ 10^{-5}T + 0.99998)$$
$$\text{(Eq. E3.9–5)}$$

Thus, the equilibrium longitudinal magnetization is on one hand inversely proportional to temperature through the spin population factor of the Boltzmann distribution, and on the other hand is a polynomial of temperature through the physical density of water. Nevertheless, as seen from the small polynomial coefficients, the dominant temperature dependence is $T^{-1}$ producing a change of approximately 0.29%/C° in the experimental range of 37 C° $-80$ C° (Chen, Daniel, & Pauly, 2006).

   As discussed in section E3.1, qMRI of the proton density is challenging for two reasons; first it does not lend itself to differential weighting and second, maps of $T_1$ and $T_2$ are needed for removing the residual $T_1$- and $T_2$-weightings from the generated pixel values. Furthermore, since the relaxation times are temperature dependent, these dependences can be confounding factors (Gultekin & Gore, 2005).

### E3.9.iii. *Temperature Mapping via Diffusion Coefficient*

The diffusion coefficient is a function of temperature through the translational correlation time. Its temperature can be modeled (Simpson & Carr,

1958) in terms of the activation $E_a$ energy of the diffusional molecular process, specifically:

$$D(T) \approx D^\infty \exp\left(-\frac{E_a}{k_B T}\right) \qquad \text{(Eq. E3.9–6)}$$

By assuming that the activation energy varies very slowly with temperature, we can differentiate the equation above with respect to temperature, leading to:

$$\frac{dD}{D} \approx \left(\frac{E_a}{k_B T^2}\right) dT \qquad \text{(Eq. E3.9–7)}$$

This can be integrated (Le Bihan, Delannoy, & Levin, 1989) resulting in the following temperature difference mapping equation:

$$(T - T_0)_{(i,j,sl)} \approx \left(\frac{k_B T_0^2}{E_a}\right)\left(\frac{D - D_0}{D_0}\right)_{(i,j,sl)} \qquad \text{(Eq. E3.9–8)}$$

between two diffusion coefficient maps generated from diffusion weighted images acquired at two different temperatures of $T$ and $T_0$ respectively. This is an approximate equation valid in the limit of small temperature variations relative to the reference image, *i.e.* $(T-T_0) << T_0$.

### E3.9.iv. *Temperature Mapping via $T_1$*

The relaxation times are increasing functions of temperature *via* the rotational and translational correlation times and have been shown to vary in gels (Parker, 1984) and in tissue *in vivo* animal study (Youl, Hawkins, Morris, DuBoulay, & Tofts, 1992). For biological tissue, within the temperature range of interest, $T_1$ is proportional to temperature to a good approximation (Young, Hand, Oatridge, & Prior, 1994).

For small temperature ranges, a Taylor series expansion of this equation leads to linear dependence of $T_1$ increasing with increasing temperature, which in turn can be transformed into a temperature difference mapping equation:

$$(T - T_{ref})_{(i,j,sl)} \approx \frac{(T_1(T) - T_1(T_{ref}))_{(i,j,sl)}}{m} \qquad \text{(Eq. E3.9–9)}$$

where the coefficient m defined by:

$$m \equiv \frac{dT_1}{dT} \qquad \text{(Eq. E3.9–10)}$$

can be tissue dependent and in general are not known a priori. Hence, from two $T_1$ maps generated from scans performed at different temperatures, a relative temperature map can be generated if m is assumed tissue independent and therefore position independent. Besides the need for assuming that the thermal coefficient, which may degrade the temperature measurement accuracy, the quantification of temperature *via* $T_1$ is difficult because the physiologic response of the living tissue to heat can seriously affect the quantification (Young, *et al.*, 1994). Nonlinear effects can also occur if the tissue properties change, due to coagulation, which has been found in *ex vivo* tissue to occur at temperatures as low as 43°C (Peller *et al.*, 2002). Recent work (Bos, Lepetit-Coiffé, Quesson, & Moonen, 2005) has demonstrated however, the possibility of combining $T_1$ and temperature mapping in near real time using a Look-Locker pulse sequence whereby the modulus images are used for $T_1$ mapping and the phase images for temperature mapping based on PRF shift.

### E3.10. *Dynamic Differential Weighting*

Dynamic differential weighting presents a different qMRI paradigm whereby image datasets are acquired at different time points in synchronicity to a bolus passage experiment in which a contrast material bolus passes through the voxels of the imaging volume. Such contrast bolus may be created exogenously by injecting an MR contrast material — *e.g.* Gadopentetic acid in the form of gadopentetate dimeglumine Gd-DTPA -- or created endogenously by saturating the magnetization of the arterial blood upstream from the imaging volume –arterial spin labeling (ASL)--.

### E3.10.i. *Time Dependent Pixel Value Equation*

When images are acquired at sufficiently high speed relative to the transit speed of the bolus, the imaging experiment effectively tracks the bolus passage and therefore these techniques are referred to as bolus tracking techniques. The final product of a bolus tracking experiment is a four-dimensional dataset and the corresponding pixel values are labeled by four indices, *i.e.*

$$
\begin{aligned}
&\mathrm{pv}_{(i,j,sl)}\left(t_{Acq\_\lambda}\right) \\
&= \Gamma_{(i,j,sl)}\mathrm{PD}_{(i,j,sl)}^{(A)}\mathrm{PSw}(\cdots q\mathrm{CV\_X}\cdots\|\cdots q\mathrm{MRI\_par}_{(i,j,sl)}(t_{Acq\_\lambda})\cdots)
\end{aligned}
$$

$$\text{(Eq. E3.10--1)}$$

where $t_{Acq-\lambda}$ the time of acquisition the $\lambda^{th}$ dataset and the main qMRI parameters causing the observed pixel value temporal changes are the relaxation times: $T_2^*$, $T_2$, and $T_1$ for dCE-Gd-DTPA bolus tracking experiments and $T_1$ for ASL bolus tracking experiments. Accordingly, bolus tracking time curves can be generated for every pixel location and these are related to the local concentration of the contrast material as a function of time. The main qMRI effects of bolus material is to alter the local relaxation times and one fundamental assumption used for analyzing bolus tracking datasets is that the time dependent concentration of the tracer are proportional to the instatntaneous change in tissue relaxation, *i.e.*

$$C_{(i,j,sl)}(t_{Acq-\lambda})\alpha \begin{cases} \Delta R^{(A)}_{1,(i,j,sl)}(t_{Acq-\lambda}) & \longleftrightarrow & T_1 - \text{weighted} \\ \Delta R^{(A)}_{2,(i,j,sl)}(t_{Acq-\lambda}) & \longleftrightarrow & T_2 - \text{weighted} \\ \Delta R^{*(A)}_{2,(i,j,sl)}(t_{Acq-\lambda}) & \longleftrightarrow & T_2^* - \text{weighted} \end{cases}$$

$$\text{(Eq. E3.10–2)}$$

For example, when tracking the passage of a Gd-DTPA bolus with a fast $T_2^*$-weighted acquisition, the magnetic susceptibility caused $R_2^*$ changes, which are dynamic, are given approximately by:

$$\Delta R^{*(A)}_{2,(i,j,sl)}(t_{Acq-\lambda}) \equiv R^{*(A)}_{2,(i,j,sl)}(t_{Acq-\lambda}) - R^{*(A)}_{2,(i,j,sl)}(0)$$

$$\cong \frac{4}{3}\pi\gamma\chi^m B_0\, BVC^{blood}_{(i,j,sl)}(t_{Acq-\lambda}) \quad \text{(Eq. E3.10–3)}$$

where $R^{*(A)}_{2,(i,j,sl)}(0)$ is the baseline value of the relaxation rate obtained by scanning before injecting the bolus, $\chi^m$ is the magnetic susceptibility of the contrast agent, BV is the blood volume fraction (%), $C^{blood}_{(i,j,sl)}(t_{Acq-\lambda})$ is the contrast agent concentration on blood, and $BVC^{blood}_{(i,j,sl)}(t_{Acq-\lambda})$ is the contrast agent concentration in tissue. The formula above is valid in the limit of high contrast agent concentration and long echo times, such as the ones typically used for dCE-Gd-DTPA bolus tracking $T_2^*$ experiments.

Similarly, when tracking a bolus with a fast $T_1$-weighted acquisition, the $T_1$ changes are given by the standard relaxivity equation:

$$\Delta R^{(A)}_{1,(i,j,sl)}(t_{Acq-\lambda}) \equiv R^{(A)}_{1,(i,j,sl)}(t_{Acq-\lambda}) - R^{(A)}_{1,(i,j,sl)}(0) = r_1 C^{tissue}_{(i,j,sl)}(t_{Acq-\lambda})$$

$$\text{(Eq. E3.10–4)}$$

where $r_1$ denotes the relaxivity of the contrast agent in units of L $mmol^{-1} s^{-1}$; numerical values for various magnetic field strengths have been reported for various commercially available MRI contrast media (Rohrer, Bauer, Mintorovitch, Requardt, & Weinmann, 2005).

From a pure theoretical qMRI standpoint, relating the measured relaxation times to concentration is a valid image processing endpoint. However, these concentrations are not directly indicative of tissue integrity and/or tissue viability, and therefore one of the main goals of dCE-MRI is the determination of regional perfusion and related hemodynamic parameters. Perfusion parameters are computed on a pixel-by-pixel basis from the concentration time curves using dilution theory as applied to the specific organ and contrast agent (tracer) used.

Because of the brain blood barrier, cerebral perfusion parameters derived from intravascular contrast agents such as Gd-DTPA are computed with a mathematical model that takes into account the fact that the contrast agent does not diffuse into the extravascular spaces. On the other hand, cerebral perfusion can be assessed with spin labeled water with a technique termed arterial spin labeling (ASL) (Williams, Detre, Leigh, & Koretsky, 1992); in this case, the perfusion algorithm is different and incorporates the fact that endogenous labeled water is a diffusable tracer. For other organs such as the liver and kidneys, where the brain blood barrier has no equivalent, the perfusion parameters are derived from the concentration time curves (PS Tofts *et al.*, 1999) with appropriate hemodynamic models that are the specific blood supply pattern, which may incorporate more than one input. For example, a dual-input one-compartment model can be used to reflect the fact that the liver has a dual blood supply from both the aorta (through the hepatic artery) and the portal vein, and therefore has two inflow rate constants (Materne *et al.*, 2002).

E3.10.ii. *Tracer Kinetics: non-diffusable tracers*

Important hemodynamic parameters are cerebral blood volume (CBV), cerebral blood flow (CBF), and mean transit time (MTT).

The CBV for a given voxel is defined by (Barbier, Lamalle, & Décorps, 2001):

$$rCBV_{(i,j,sl)} = \frac{\text{Volume of blood in a voxel}}{\text{Mass of the voxel}} \quad \text{(Eq. E3.10–5)}$$

and is usually expressed in units of microliters per gram ($\mu L\ g^{-1}$); typical values for the adult human brain are about $50\,\mu L\ g^{-1}$.

In turn, the voxel CBF is defined by:

$$\text{rCBF}_{(i,j,sl)} = \frac{\text{Net blood flow through a voxel}}{\text{Mass of the voxel}} \qquad \text{(Eq. E3.10–6)}$$

and is usually expressed in units of microliters per second per gram ($\mu L\ s^{-1}\ g^{-1}$); typical values for the adult brain obtained with non-MR methods are about $10\ \mu L\ s^{-1}\ g^{-1}$.

The central equations for the determination of the regional CBV and CBF stem from (Østergaard, Weisskoff, Chesler, Gyldensted, & Rosen, 1996) a convolution equation expressing the concentration at time t in terms of an arterial input function, which is also a function of time an characterizes the blood supply to the volume of interest. Accordingly:

$$C_{v,(i,j,sl)}(t) = C_{a,(i,j,sl)}(t) * h_{(i,j,sl)}(t) \equiv \int_{-\infty}^{t} C_{a(i,j,sl)}(\tau) * h_{(i,j,sl)}(t - \tau)d\tau$$

$$\text{(Eq. E3.10–7)}$$

where h denotes the probability density function of the transit times. Hence, the mean transit time of a given voxel is given by:

$$\text{MTT}_{(i,j,sl)} = \frac{\int_{-\infty}^{+\infty} \tau h_{(i,j,sl)}(\tau)d\tau}{\int_{-\infty}^{+\infty} h_{(i,j,sl)}(\tau)d\tau} \qquad \text{(Eq. E3.10–8)}$$

The amount of intravascular tracer in the voxel is determined with the following equation:

$$\text{CBV}_{(i,j,sl)} = \frac{\int_{-\infty}^{+\infty} C_{(i,j,sl)}(\tau)d\tau}{\int_{-\infty}^{+\infty} C_{a}(\tau)d\tau} \qquad \text{(Eq. E3.10–9)}$$

With the definitions of MTT and CBV above and invoking the central volume theorem, it can be shown (Meier & Zierler, 1954) that these two hemodynamic parameters are related by:

$$\text{MTT}_{(i,j,sl)} = \frac{\text{CBV}_{(i,j,sl)}}{F_{(i,j,sl)}} \qquad \text{(Eq. E3.10–10)}$$

where F is the tissue flow, or perfusion and is expressed in units of ($\mu L\ s^{-1}\ g^{-1}$).

A central quantity in bolus-passage experiments is the fraction of injected tracer still present in the vasculature at time t. This is described by the residue function R(t), as defined by:

$$R_{(i,j,sl)}(t) \equiv 1 - \int_0^t h_{(i,j,sl)}(\tau)d\tau \qquad \text{(Eq. E3.10-11)}$$

which is a positive and decreasing function of time, starting from $R(0)=1$. The concentration at a given voxel can now be written in terms of flow and the convolution of the arterial input function with the residue function.

$$C_{(i,j,sl)}(t) = F_{(i,j,sl)}C_{a,(i,j,sl)}(t) * R_{(i,j,sl)}(t) \qquad \text{(Eq. E3.10-12)}$$

This is the central equation for computing flow maps with non-diffusable tracers (Østergaard, *et al.*, 1996). It predicts that the initial height of the deconvolved concentration time curve equals the flow. These investigators note that the arterial input function in this equation may undergo dispersion during its passage from the point of measurement to more peripheral tissue thus underlining the importance of measuring the arterial input values close to the observed tissue to avoid dispersion.

This convolution equation is not straightforward to solve because the residue function is in general unknown and depends on the structure of the intravoxel vasculature. From a mathematical standpoint, this is an inverse type problem where integral equations are solved with respect to an unknown kernel, the residue function in this particular case. The approaches to deconvolve such inverse problems follow in two main categories (Østergaard, *et al.*, 1996): model dependent techniques in which an specific analytical form for the residue function is assumed in a parametric form and model independent techniques where both flow and the shape of the residue function are determined from the experimental curves by nonparametric deconvolution. Further details into these techniques are beyond the scope of this book and the reader is referred to the above-cited papers (Barbier, *et al.*, 2001; Meier & Zierler, 1954; Østergaard, *et al.*, 1996) and references therein for a more in depth discussion.

### E3.10.iii. *Tracer Kinetics: diffusable tracer*

The rate of accumulation and washout of an extracellular contrast agent in the extravascular extracellular space (ESS) can be modeled by a first order differential equation of the form (PS Tofts, *et al.*, 1999):

$$\frac{dC_e}{dt} = K^{trans}\left(C_p(t) - \frac{C_e(t)}{v_e}\right) \qquad \text{(Eq. E3.10-13)}$$

where $v_e$ is the extravascular extracellular space volume per unit volume, $C_e(t)$ is the concentration of the contrast agent in $v_e$ and $C_p(t)$ is the concentration of contrast agent in arterial blood plasma, and $K^{tran}$ is the volume transfer constant between blood plasma and ESS, often referred in abbreviate form as the transfer constant.

The physiologic interpretation of the transfer constant relates to the balance between capillary wall permeability and tissue blood flow (PS Tofts, et al., 1999). In high-permeability limit, where flux across the endothelium is flow limited, the transfer constant is equal to the blood plasma flow per unit volume of tissue whereas in the low permeability limit, where tracer flux is permeability limited, the transfer constant is equal to the permeability surface area product between blood plasma and the EES. In the case in which the tracer uptake is limited by both blood flow and capillary wall permeability, one needs to consider the extraction ratio E, which is the fractional reduction in capillary blood concentration as it passes through tissue.

$$E = \frac{C_a - C_v}{C_a}$$  (Eq. E3.10–14)

Altogether, we may summarize these regimes with the following equation:

$$K^{trans} = \begin{cases} PS\rho & \leftrightarrow & PS << F \\ EF\rho(1 - Hct) & \leftrightarrow & \text{mixed flow and Ps limited regime} \\ F\rho(1 - Hct) & \leftrightarrow & PS >> F \end{cases}$$

(Eq. E3.10–15)

where $\rho$ is the tissue density (g ml$^{-1}$), PS is the permeability surface area product per unit mass of tissue, F is flow as defined in the previous section, and Hct is the hematocrit --or volume fraction occupied by red blood cells, typically about 40%--.

These equations constitute the generalized kinetic model for diffusable tracers in tissue. Developments that are more specific are beyond the scope of this book and the reader is referred to the review paper (PS Tofts, et al., 1999) and references therein.

### E3.10.iv. Arterial Spin Labeling (ASL)

The general principles and basic methodologies of arterial spin labeling were laid out by Detre et al. (Detre, Leigh, Williams, & Koretsky, 1992) and by Kwong et al. (Kwong et al., 1992). The basic ASL experiment consists in applying a water-labeling slice selective rf pulse acting on spins upstream from the imaging slice positioned at a certain distance downstream. Arterial water can be labeled *via* saturation or inversion, the latter being preferable because it leads to a larger effect (Roberts, Detre, Bolinger, Insko, & Leigh, 1994). The theoretical underpinnings of ASL stem from solutions to a modified longitudinal magnetization Bloch equation with added terms modeling the flow of blood, specifically (Detre, *et al.*, 1992):

$$\frac{dM_z}{dt} = \frac{(M_z^{(eq)} - M_z)}{T_1} + rCBF(M_z^{(arterial)} - M_z^{(venous)})$$

(Eq. E3.10–16)

At equilibrium, the relationship between the blood and the tissue longitudinal magnetizations involves a parameter termed blood-brain partition coefficient ($\lambda$) *via*:

$$M_z^{(eq\_arterial)} = M_z^{(eq\_venous)} = \frac{M_z^{(eq)}}{\lambda} \qquad \text{(Eq. E3.10–17)}$$

The blood-brain partition coefficient is the difference between the water concentration in blood and brain tissue.

Assuming a full exchange between blood and tissue water, the venous magnetization is equal to the tissue magnetization, *i.e.*

$$M_z^{(venous)}(t) = \frac{M_z(t)}{\lambda} \qquad \text{(Eq. E3.10–18)}$$

Furthermore, in the absence of upstream arterial saturation, and again assuming a full exchange between blood and tissue water, the arterial magnetization can be assumed constant, hence:

$$M_z^{(arterial)}(t) = M_z^{(arterial)}(0) = \frac{M_z^{(eq)}}{\lambda} \qquad \text{(Eq. E3.10–19)}$$

On the other hand, after the arterial magnetization has been saturated upstream (proximal saturation), the arterial longitudinal magnetization can be assumed null for the duration of an imaging experiment, $M_z^{(arterial)}(t)=0$. Combining, these two possible cases in a single expression:

$$M_z^{(arterial)}(t) = \begin{cases} 0 & \leftrightarrow \quad \text{proximal saturation} \\ \dfrac{M_z^{(eq)}}{\lambda} & \leftrightarrow \qquad \text{control} \end{cases} \qquad \text{(Eq. E3.10–20)}$$

Hence, the longitudinal Bloch equation,

$$\frac{dM_z}{dt} = \frac{(M_z^{(eq)} - M_z)}{T_1} + rCBF \left( M_z^{(arterial)} - \frac{M_z}{\lambda} \right) \quad \text{(Eq. E3.10–21)}$$

can be integrated. Here we are interested in solving it for two imaging experiments performed sequentially, with and without the proximal presaturation of arterial blood. In the case of fast gradient-echo (FLASH) sequences, the solution to Eq. E3.10-21 with a proximal presaturation pulse applied upstream is:

$$M_z^{ss}(TR) = \frac{M_z^{(eq)}}{\left(1 + rCBF \frac{T_1}{\lambda}\right)} \left( \frac{1 - \exp\left(-\frac{TR}{T_1^{App}}\right)}{1 - \cos(FA) \exp\left(-\frac{TR}{T_1^{App}}\right)} \right)$$

$$\text{(Eq. E3.10–22)}$$

and without proximal presaturation:

$$M_z^{control}(TR) = M_z^{(eq)} \left( \frac{1 - \exp\left(-\frac{TR}{T_1^{App}}\right)}{1 - \cos(FA) \exp\left(-\frac{TR}{T_1^{App}}\right)} \right) \quad \text{(Eq. E3.10–23)}$$

where the apparent longitudinal relaxation has been defined as:

$$T_1^{App} \equiv \left( \frac{1}{T_1} + \frac{rCBF}{\lambda} \right)^{-1} \quad \text{(Eq. E3.10–24)}$$

From the ratio of the solutions above, we find:

$$\frac{M_z^{ss}(TR)}{M_z^{control}(TR)} = \frac{1}{\left(1 + rCBF \frac{T_1}{\lambda}\right)} \quad \text{(Eq. E3.10–25)}$$

which can be rearranged leading to:

$$rCBF = \frac{\lambda}{T_1} \left( \frac{M_z^{control}(TR)}{M_z^{ss}(TR)} - 1 \right) \quad \text{(Eq. E3.10–26)}$$

and, in a final step by combining this with Eq. E3.12-22, rCBF can be expressed in terms of measurable quantities, specifically:

$$rCBF = \frac{\lambda}{T_1^{App}} \left( \frac{M_z^{control}(TR) - M_z^{ss}(TR)}{M_z^{control}(TR)} \right) \quad \text{(Eq. E3.10–27)}$$

This basic mapping equation serves as the foundation for ASL techniques. rCBF can be mapped with an imaging experiment consisting of two ASL acquisitions and a map of the apparent $T_1$. In practice, this result needs to be modified to include a parameter $\alpha$ representing the effective degree of arterial inversion (Roberts, *et al.*, 1994) leading to:

$$rCBF_{(i,j,sl)} = \frac{\lambda}{2\alpha'T_{1,(i,j,sl)}^{App}} \left( \frac{M_{z,(i,j,sl)}^{control}(TR) - M_{z,(i,j,sl)}^{ss}(TR)}{M_{z,(i,j,sl)}^{control}(TR)} \right)$$

(Eq. E3.10–28)

where the pixel indices are shown explicitly. The effective degree of inversion at the imaging slice is in turn dependent on the $T_1$ of arterial blood and on pulse sequence parameters as follows:

$$\alpha' = \exp\left(-\frac{\Delta}{T_1^{arterial}}\right)\left(\frac{T_{inv}}{TR}\right)\alpha \qquad (Eq.\ E3.10–29)$$

where $\Delta$ and $T_{inv}$ are the time of travel of arterial blood from the inversion plane to the imaging slice and the duration of the inversion pulse respectively. Furthermore, $\alpha$ denotes the degree of inversion at the inversion plane.

In the context of this book, ASL is a differentially weighted qMRI technique, with which the targeted qMRI parameter is perfusion (rCBF or in general f), which is mapped *via* the difference of the two directly acquired images, with and without perfusion weighting.

One difficulty associated with ASL techniques is the concomitant MT-weighting resulting from applying the rf labeling pulse, which introduces an MT bias in the labeled image relative to the control image. Hence, it is necessary to MT-weight the control image to the same degree. This can be accomplished by acquiring the control image with a downstream symmetrically positioned (non-) labeling rf pulse. This is a single slice technique and therefore not ideally suited for clinical applications. A multisection technique has been described (Alsop & Detre, 1998) whereby the application of an amplitude-modulated form of the labeling rf irradiation effectively produces a double inversion near the labeling plane. Double inversion produces no net effect, so spins are not labeled by the amplitude-modulated in the control image. Furthermore, because the average power and center frequency of the amplitude- modulated control are identical to those of the labeling rf irradiation, the off-resonance effects of the control can be nearly identical to those of the labeling (Alsop & Detre, 1998).

# References

Ababneh, Z., Beloeil, H., Berde, C., Ababneh, A., Maier, S., & Mulkern, R. (2009). In vivo lipid diffusion coefficient measurements in rat bone marrow. *Magnetic resonance imaging, 27*(6), 859–864.

Aletras, A., Ding, S., Balaban, R., & Wen, H. (1999). DENSE: displacement encoding with stimulated echoes in cardiac functional MRI. *Journal of Magnetic Resonance, 137*(1), 247–252.

Alsop, D., & Detre, J. (1998). Multisection cerebral blood flow MR imaging with continuous arterial spin labeling. *Radiology, 208*(2), 410.

An, H., & Lin, W. (2002). Cerebral oxygen extraction fraction and cerebral venous blood volume measurements using MRI: effects of magnetic field variation. *Magnetic Resonance in Medicine, 47*(5), 958–966.

Axel, L., & Dougherty, L. (1989a). Heart wall motion: improved method of spatial modulation of magnetization for MR imaging. *Radiology, 172*(2), 349.

Axel, L., & Dougherty, L. (1989b). MR imaging of motion with spatial modulation of magnetization. *Radiology, 171*(3), 841.

Barbier, E., Lamalle, L., & Décorps, M. (2001). Methodology of brain perfusion imaging. *Journal of Magnetic Resonance Imaging, 13*(4), 496–520.

Basser, P., & Pierpaoli, C. (1998). A simplified method to measure the diffusion tensor from seven MR images. *Magnetic Resonance in Medicine, 39*(6), 928–934.

Basser, P. J., Mattiello, J., & LeBihan, D. (1994). Estimation of the effective self-diffusion tensor from the NMR spin echo. *J Magn Reson B, 103*(3), 247–254.

Baudrexel, S., Volz, S., Preibisch, C., Klein, J., Steinmetz, H., Hilker, R., et al. (2009). Rapid single-scan T2*-mapping using exponential excitation pulses and image-based correction for linear background gradients. *Magnetic Resonance in Medicine, 62*(1), 263–268.

Bluml, S., Schad, L. R., Stepanow, B., & Lorenz, W. J. (1993). Spin-lattice relaxation time measurement by means of a TurboFLASH technique. *Magn Reson Med, 30*(3), 289–295.

Bos, C., Lepetit-Coiffé, M., Quesson, B., & Moonen, C. (2005). Simultaneous monitoring of temperature and T1: methods and preliminary results of application to drug delivery using thermosensitive liposomes. *Magnetic Resonance in Medicine, 54*(4), 1020–1024.

Brix, G., Schad, L. R., Deimling, M., & Lorenz, W. J. (1990). Fast and precise T1 imaging using a TOMROP sequence. *Magn Reson Imaging, 8*(4), 351–356.

Chebrolu, V., Hines, C., Yu, H., Pineda, A., Shimakawa, A., McKenzie, C., et al. (2010). Independent estimation of T2* for water and fat for improved accuracy of fat quantification. *Magnetic Resonance in Medicine, 63*(4), 849–857.

Chen, J., Daniel, B., & Pauly, K. (2006). Investigation of proton density for measuring tissue temperature. *Journal of Magnetic Resonance Imaging, 23*(3), 430–434.

Cheng, H. L., & Wright, G. A. (2006). Rapid high-resolution T(1) mapping by variable flip angles: accurate and precise measurements in the presence of radiofrequency field inhomogeneity. *Magn Reson Med, 55*(3), 566–574.

Constable, R. (1995). Functional MR imaging using gradient-echo echo-planar imaging in the presence of large static field inhomogeneities. *Journal of magnetic resonance imaging: JMRI, 5*(6), 746.

De Poorter, J. (1995). Noninvasive MRI thermometry with the proton resonance frequency method: study of susceptibility effects. *Magnetic Resonance in Medicine, 34*(3), 359–367.

De Poorter, J., De Wagter, C., De Deene, Y., Thomsen, C., Ståhlberg, F., & Achten, E. (1995). Noninvasive MRI thermometry with the proton resonance frequency (PRF) method: in vivo results in human muscle. *Magnetic Resonance in Medicine, 33*(1), 74–81.

Deichmann, R., Adolf, H., Kuchenbrod, E., Noth, U., Schwarzbauer, C., & Haase, A. (1995). Compensation of diffusion effects in T2 measurements. *Magn Reson Med, 33*(1), 113–115.

Deichmann, R., Hahn, D., & Haase, A. (1999). Fast T1 mapping on a whole-body scanner. *Magn Reson Med, 42*(1), 206–209.

Deoni, S. C. (2007). High-resolution T1 mapping of the brain at 3T with driven equilibrium single pulse observation of T1 with high-speed incorporation of RF field inhomogeneities (DESPOT1-HIFI). *J Magn Reson Imaging, 26*(4), 1106–1111.

Deoni, S. C., Rutt, B. K., & Peters, T. M. (2003). Rapid combined T1 and T2 mapping using gradient recalled acquisition in the steady state. *Magn Reson Med, 49*(3), 515–526.

Deoni, S. C., Ward, H. A., Peters, T. M., & Rutt, B. K. (2004). Rapid T2 estimation with phase-cycled variable nutation steady-state free precession. *Magn Reson Med, 52*(2), 435–439.

Detre, J., Leigh, J., Williams, D., & Koretsky, A. (1992). Perfusion imaging. *Magnetic Resonance in Medicine, 23*(1), 37–45.

Dixon, W., Engels, H., Castillo, M., & Sardashti, M. (1990). Incidental magnetization transfer contrast in standard multislice imaging. *Magnetic resonance imaging, 8*(4), 417–422.

Dixon, W. T. (1984). Simple proton spectroscopic imaging. *Radiology, 153*(1), 189–194.

Does, M., & Gore, J. (2000). Complications of nonlinear echo time spacing for measurement of T2. *NMR in Biomedicine, 13*(1), 1–7.

Does, M., Parsons, E., & Gore, J. (2003). Oscillating gradient measurements of water diffusion in normal and globally ischemic rat brain. *Magnetic Resonance in Medicine, 49*(2), 206–215.

Edzes, H., & Samulski, E. (1977). Cross relaxation and spin diffusion in the proton NMR or hydrated collagen. *Nature, 265*(5594), 521.

Eng, J., Ceckler, T. L., & Balaban, R. S. (1991). Quantitative 1H magnetization transfer imaging in vivo. *Magn Reson Med, 17*(2), 304–314.

Fernández-Seara, M., & Wehrli, F. (2000). Postprocessing technique to correct for background gradients in image-based R* 2 measurements. *Magnetic Resonance in Medicine, 44*(3), 358–366.

Frahm, J., Merboldt, K., & Hänicke, W. (1988). Direct FLASH MR imaging of magnetic field inhomogeneities by gradient compensation. *Magnetic* resonance in medicine: official journal of the Society of Magnetic Resonance in Medicine/Society of Magnetic Resonance in Medicine, 6(4), 474.

Glover, G. (1991). Multipoint Dixon technique for water and fat proton and susceptibility imaging. *Journal of Magnetic Resonance Imaging, 1*(5), 521–530.

Glover, G., & Schneider, E. (1991). Three-point Dixon technique for true water/fat decomposition with B0 inhomogeneity correction. *Magnetic Resonance in Medicine, 18*(2), 371–383.

Gowland, P., & Mansfield, P. (1993). Accurate measurement of T1 in vivo in less than 3 seconds using echo-planar imaging. *Magn Reson Med, 30*(3), 351–354.

Gowland, P. A., & Leach, M. O. (1992). Fast and accurate measurements of T1 using a multi-readout single inversion-recovery sequence. *Magn Reson Med, 26*(1), 79–88.

Grad, J., Mendelson, D., Hyder, F., & Bryant, R. (1991). Applications of nuclear magnetic cross relaxation spectroscopy to tissues. *Magnetic Resonance in Medicine, 17*(2), 452–459.

Graham, S. J., & Henkelman, R. M. (1997). Understanding pulsed magnetization transfer. *J Magn Reson Imaging, 7*(5), 903–912.

Gudbjartsson, H., Maier, S. E., Mulkern, R. V., Morocz, I. A., Patz, S., & Jolesz, F. A. (1996). Line scan diffusion imaging. *Magn Reson Med, 36*(4), 509–519.

Gultekin, D., & Gore, J. (2005). Temperature dependence of nuclear magnetization and relaxation. *Journal of Magnetic Resonance, 172*(1), 133–141.

Haase, A. (1990). Snapshot FLASH MRI. Applications to T1, T2, and chemical-shift imaging. *Magn Reson Med, 13*(1), 77–89.

Henderson, E., McKinnon, G., Lee, T. Y., & Rutt, B. K. (1999). A fast 3D look-locker method for volumetric T1 mapping. *Magn Reson Imaging, 17*(8), 1163–1171.

Henkelman, R. M., Huang, X., Xiang, Q. S., Stanisz, G. J., Swanson, S. D., & Bronskill, M. J. (1993). Quantitative interpretation of magnetization transfer. *Magn Reson Med, 29*(6), 759–766.

Hilaire, L., Wehrli, F., & Song, H. (2000). High-speed spectroscopic imaging for cancellous bone marrow R2* mapping and lipid quantification. *Magnetic resonance imaging, 18*(7), 777–786.

Hindman, J. (1966). Proton resonance shift of water in the gas and liquid states. *The Journal of Chemical Physics, 44*, 4582.

Hindman, J., Svirmickas, A., & Wood, M. (1973). Relaxation processes in water. A study of the proton spin lattice relaxation time. *The Journal of Chemical Physics, 59*, 1517.

Holdsworth, S. J., Skare, S., Newbould, R. D., & Bammer, R. (2009). Robust GRAPPA-accelerated diffusion-weighted readout-segmented (RS)-EPI. *Magn Reson Med, 62*(6), 1629–1640.

Ishihara, Y., Calderon, A., Watanabe, H., Okamoto, K., Suzuki, Y., Kuroda, K., et al. (1995). A precise and fast temperature mapping using water proton chemical shift. *Magnetic Resonance in Medicine, 34*(6), 814–823.

Kay, I., & Henkelman, R. M. (1991). Practical implementation and optimization of one-shot T1 imaging. *Magn Reson Med, 22*(2), 414–424.

Kubicki, M., Maier, S., Westin, C., Mamata, H., Ersner-Hershfield, H., Estepar, R., *et al.* (2004). Comparison of single-shot echo-planar and line scan protocols for diffusion tensor imaging. *Academic radiology, 11*(2), 224.

Kwong, K., Belliveau, J., Chesler, D., Goldberg, I., Weisskoff, R., Poncelet, B., *et al.* (1992). Dynamic magnetic resonance imaging of human brain activity during primary sensory stimulation. *Proceedings of the National Academy of Sciences, 89*(12), 5675.

Le Bihan, D. (2008). Intravoxel Incoherent Motion Perfusion MR Imaging: A Wake-Up Call1. *Radiology, 249*(3), 748.

Le Bihan, D., & Breton, E. (1985). Imagerie de diffusion in vivo par résonance magnétique nucléaire= In vivo magnetic resonance imaging of diffusion. *Comptes rendus de l'Académie des sciences. Série 2, Mécanique, Physique, Chimie, Sciences de l'univers, Sciences de la Terre, 301*(15), 1109–1112.

Le Bihan, D., Breton, E., Lallemand, D., Aubin, M., Vignaud, J., & Laval-Jeantet, M. (1988). Separation of diffusion and perfusion in intravoxel incoherent motion MR imaging. *Radiology, 168*(2), 497.

Le Bihan, D., Delannoy, J., & Levin, R. (1989). Temperature mapping with MR imaging of molecular diffusion: application to hyperthermia. *Radiology, 171*(3), 853.

Le Bihan, D., Poupon, C., Amadon, A., & Lethimonnier, F. (2006). Artifacts and pitfalls in diffusion MRI. *Journal of Magnetic Resonance Imaging, 24*(3), 478–488.

Lehnert, A., Machann, J., Helms, G., Claussen, C., & Schick, F. (2004). Diffusion characteristics of large molecules assessed by proton MRS on a whole-body MR system. *Magnetic resonance imaging, 22*(1), 39–46.

Li, J. G., Graham, S. J., & Henkelman, R. M. (1997). A flexible magnetization transfer line shape derived from tissue experimental data. *Magn Reson Med, 37*(6), 866–871.

Liu, C. Y., McKenzie, C. A., Yu, H., Brittain, J. H., & Reeder, S. B. (2007). Fat quantification with IDEAL gradient echo imaging: correction of bias from T(1) and noise. *Magn Reson Med, 58*(2), 354–364.

Look, D. C., & Locker, D. R. (1970). Time Saving in Measurement of NMR and EPR Relaxation Times. *Review of Scientific Instruments, 41*(2), 250–251.

Luciani, A., Vignaud, A., Cavet, M., Tran Van Nhieu, J., Mallat, A., Ruel, L., *et al.* (2008). Liver Cirrhosis: Intravoxel Incoherent Motion MR Imaging— Pilot Study1. *Radiology, 249*(3), 891.

Majumdar, S., & Gore, J. C. (1987). Effects of selective pulses on the measurement of T2 and apparent diffusion in multiecho MRI. *Magn Reson Med, 4*(2), 120–128.

Majumdar, S., Orphanoudakis, S. C., Gmitro, A., O'Donnell, M., & Gore, J. C. (1986a). Errors in the measurements of T2 using multiple-echo MRI techniques. I. Effects of radiofrequency pulse imperfections. *Magn Reson Med, 3*(3), 397–417.

Majumdar, S., Orphanoudakis, S. C., Gmitro, A., O'Donnell, M., & Gore, J. C. (1986b). Errors in the measurements of T2 using multiple-echo MRI techniques. II. Effects of static field inhomogeneity. *Magn Reson Med, 3*(4), 562–574.

Materne, R., Smith, A., Peeters, F., Dehoux, J., Keyeux, A., Horsmans, Y., *et al.* (2002). Assessment of hepatic perfusion parameters with dynamic MRI. *Magnetic Resonance in Medicine, 47*(1), 135–142.

Mattiello, J., Basser, P. J., & Le Bihan, D. (1997). The b matrix in diffusion tensor echo-planar imaging. *Magn Reson Med, 37*(2), 292–300.

Meiboom, S., & Gill, D. (1958). Modified spin echo method for measuring nuclear relaxation times. *Review of Scientific Instruments, 29*, 688.

Meier, P., & Zierler, K. (1954). On the theory of the indicator-dilution method for measurement of blood flow and volume. *Appl Physiol, 6*, 731–744.

Melki, P. S., & Mulkern, R. V. (1992). Magnetization transfer effects in multislice RARE sequences. *Magn Reson Med, 24*(1), 189–195.

Melki, P. S., Mulkern, R. V., Panych, L. P., & Jolesz, F. A. (1991). Comparing the FAISE method with conventional dual-echo sequences. *J Magn Reson Imaging, 1*(3), 319–326.

Merboldt, K. D., Hanicke, W., & Frahm, J. (1985). Self-diffusion NMR imaging using stimulated echoes. *Journal of Magnetic Resonance (1969), 64*(3), 479–486.

Miller, A., & Joseph, P. (1993). The use of power images to perform quantitative analysis on low SNR MR images. *Magnetic resonance imaging, 11*(7), 1051–1056.

Moran, P. (1982). A flow velocity zeugmatographic interlace for NMR imaging in humans. *Magnetic resonance imaging, 1*(4), 197–203.

Morrison, C., & Henkelman, R. M. (1995). A model for magnetization transfer in tissues. *Magn Reson Med, 33*(4), 475–482.

Morrison, C., Stanisz, G., & Henkelman, R. M. (1995). Modeling magnetization transfer for biological-like systems using a semi-solid pool with a super-Lorentzian lineshape and dipolar reservoir. *J Magn Reson B, 108*(2), 103–113.

Némethy, G., & Scheraga, H. (1962). Structure of water and hydrophobic bonding in proteins. I. A model for the thermodynamic properties of liquid water. *The Journal of Chemical Physics, 36*, 3382.

Nielsen, J., & Nayak, K. (2009). Referenceless phase velocity mapping using balanced SSFP. *Magnetic Resonance in Medicine, 61*(5), 1096–1102.

Norris, D. (2001). Implications of bulk motion for diffusion-weighted imaging experiments: effects, mechanisms, and solutions. *Journal of Magnetic Resonance Imaging, 13*(4), 486–495.

Østergaard, L., Weisskoff, R., Chesler, D., Gyldensted, C., & Rosen, B. (1996). High resolution measurement of cerebral blood flow using intravascular tracer bolus passages. Part I: Mathematical approach and statistical analysis. *Magnetic Resonance in Medicine, 36*(5), 715–725.

Parker, D. L. (1984). Applications of NMR imaging in hyperthermia: an evaluation of the potential for localized tissue heating and noninvasive temperature monitoring. *IEEE Trans Biomed Eng, 31*(1), 161–167.

Pelc, N., Bernstein, M., Shimakawa, A., & Glover, G. (1991). Encoding strategies for three-direction phase-contrast MR imaging of flow. *Journal of Magnetic Resonance Imaging, 1*(4), 405–413.

Peller, M., Reinl, H., Weigel, A., Meininger, M., Issels, R., & Reiser, M. (2002). T1 relaxation time at 0.2 Tesla for monitoring regional hyperthermia: Feasibility study in muscle and adipose tissue. *Magnetic Resonance in Medicine, 47*(6), 1194–1201.

Peters, R. D., Hinks, R. S., & Henkelman, R. M. (1998). Ex vivo tissue-type independence in proton-resonance frequency shift MR thermometry. *Magn Reson Med, 40*(3), 454–459.

Pierpaoli, C., Jezzard, P., Basser, P. J., Barnett, A., & Di Chiro, G. (1996). Diffusion tensor MR imaging of the human brain. *Radiology, 201*(3), 637–648.

Poon, C. S., & Henkelman, R. M. (1992). Practical T2 quantitation for clinical applications. *J Magn Reson Imaging, 2*(5), 541–553.

Portnoy, S., & Stanisz, G. (2007). Modeling pulsed magnetization transfer. *Magnetic Resonance in Medicine, 58*(1), 144–155.

Posse, S., Shen, Z., Kiselev, V., & Kemna, L. (2003). Single-shot T2* mapping with 3D compensation of local susceptibility gradients in multiple regions. *Neuroimage, 18*(2), 390–400.

Preibisch, C., & Deichmann, R. (2009). Influence of RF spoiling on the stability and accuracy of T1 mapping based on spoiled FLASH with varying flip angles. *Magnetic Resonance in Medicine, 61*(1), 125–135.

Press, W., Teukolsky, S., Vetterling, W., & Flannery, B. (1993). *Numerical recipes in Fortran; the art of scientific computing*: Cambridge University Press New York, NY, USA.

Ramani, A., Dalton, C., Miller, D. H., Tofts, P. S., & Barker, G. J. (2002). Precise estimate of fundamental in-vivo MT parameters in human brain in clinically feasible times. *Magn Reson Imaging, 20*(10), 721–731.

Raya, J., Dietrich, O., Horng, A., Weber, J., Reiser, M., & Glaser, C. (2010). T2 measurement in articular cartilage: Impact of the fitting method on accuracy and precision at low SNR. *Magnetic Resonance in Medicine, 63*(1), 181–193.

Reeder, S. B., Hargreaves, B. A., Yu, H., & Brittain, J. H. (2005). Homodyne reconstruction and IDEAL water-fat decomposition. *Magn Reson Med, 54*(3), 586–593.

Reeder, S. B., Pineda, A. R., Wen, Z., Shimakawa, A., Yu, H., Brittain, J. H., et al. (2005). Iterative decomposition of water and fat with echo asymmetry and least-squares estimation (IDEAL): application with fast spin-echo imaging. *Magn Reson Med, 54*(3), 636–644.

Reeder, S. B., Wen, Z., Yu, H., Pineda, A. R., Gold, G. E., Markl, M., et al. (2004). Multicoil Dixon chemical species separation with an iterative least-squares estimation method. *Magn Reson Med, 51*(1), 35–45.

Reese, T., Heid, O., Weisskoff, R., & Wedeen, V. (2003). Reduction of Eddy-Current-Induced Distortion in Diffusion MRI Using a Twice-Refocused Spin Echo. *Magnetic Resonance in Medicine, 49*, 177–182.

Ren, J., Dimitrov, I., Sherry, A., & Malloy, C. (2008). Composition of adipose tissue and marrow fat in humans by 1H NMR at 7 Tesla. *The Journal of Lipid Research, 49*(9), 2055.

Rieke, V., & Pauly, K. B. (2008a). Echo combination to reduce proton resonance frequency (PRF) thermometry errors from fat. *Journal of Magnetic Resonance Imaging, 27*(3), 673–677.

Rieke, V., & Pauly, K. B. (2008b). MR thermometry. *Journal of Magnetic Resonance Imaging, 27*(2), 376–390.

Roberts, D., Detre, J., Bolinger, L., Insko, E., & Leigh, J. (1994). Quantitative magnetic resonance imaging of human brain perfusion at 1.5 T using steady-state inversion of arterial water. *Proceedings of the National Academy of Sciences of the United States of America, 91*(1), 33.

Rohrer, M., Bauer, H., Mintorovitch, J., Requardt, M., & Weinmann, H. (2005). Comparison of magnetic properties of MRI contrast media solutions at different magnetic field strengths. *Investigative radiology, 40*(11), 715–724.

Ropele, S., Bammer, R., Stollberger, R., & Fazekas, F. (2001). T1 maps from shifted spin echoes and stimulated echoes. *Magn Reson Med, 46*(6), 1242–1245.

Ropele, S., Stollberger, R., Ebner, F., & Fazekas, F. (1999). T1 imaging using phase acquisition of composite echoes. *Magn Reson Med, 41*(2), 386–391.

Ropele, S., Stollberger, R., Kapeller, P., Hartung, H. P., & Fazekas, F. (1999). Fast multislice T(1) and T(1sat) imaging using a phase acquisition of composite echoes (PACE) technique. *Magn Reson Med, 42*(6), 1089–1097.

Santyr, G. (1993). Magnetization transfer effects in multislice MR imaging. *Magnetic resonance imaging, 11*(4), 521–532.

Schachter, M., Does, M., Anderson, A., & Gore, J. (2000). Measurements of restricted diffusion using an oscillating gradient spin-echo sequence. *Journal of Magnetic Resonance, 147*(2), 232–237.

Scheffler, K., & Hennig, J. (2001). T(1) quantification with inversion recovery TrueFISP. *Magn Reson Med, 45*(4), 720–723.

Schmitt, P., Griswold, M. A., Jakob, P. M., Kotas, M., Gulani, V., Flentje, M., et al. (2004). Inversion recovery TrueFISP: quantification of T(1), T(2), and spin density. *Magn Reson Med, 51*(4), 661–667.

Schneider, W., Bernstein, H., & Pople, J. (1958). Proton magnetic resonance chemical shift of free (gaseous) and associated (liquid) hydride molecules. *The Journal of Chemical Physics, 28*, 601.

Simpson, J., & Carr, H. (1958). Diffusion and nuclear spin relaxation in water. *Physical Review, 111*(5), 1201–1202.

Skinner, T. E., & Glover, G. H. (1997). An extended two-point Dixon algorithm for calculating separate water, fat, and B0 images. *Magn Reson Med, 37*(4), 628–630.

Sled, J., & Pike, G. (2000). Quantitative Interpretation of Magnetization Transfer in Spoiled Gradient Echo MRI Sequences* 1. *Journal of Magnetic Resonance, 145*(1), 24–36.

Sled, J. G., & Pike, G. B. (2000). Correction for B(1) and B(0) variations in quantitative T(2) measurements using MRI. *Magn Reson Med, 43*(4), 589–593.

Sled, J. G., & Pike, G. B. (2001). Quantitative imaging of magnetization transfer exchange and relaxation properties in vivo using MRI. *Magn Reson Med, 46*(5), 923–931.

Steinhoff, S., Zaitsev, M., Zilles, K., & Shah, N. J. (2001). Fast T(1) mapping with volume coverage. *Magn Reson Med, 46*(1), 131–140.

Stejskal, E. O., & Tanner, J. E. (1965). Spin Diffusion Measurements: Spin Echoes in the Presence of a Time-Dependent Field Gradient. *Journal of Chemical Physics, Vol. 42*, 288–292.

Taylor, D., & Bushell, M. (1985). The spatial mapping of translational diffusion coefficients by the NMR imaging technique. *Physics in Medicine and Biology, 30*, 345.

Tofts, P. (2003). *Quantitative MRI of the brain: measuring changes caused by disease*: John Wiley & Sons Inc.

Tofts, P., Brix, G., Buckley, D., Evelhoch, J., Henderson, E., Knopp, M., et al. (1999). Estimating kinetic parameters from dynamic contrast-enhanced T1-weighted MRI of a diffusable tracer: standardized quantities and symbols. *Journal of Magnetic Resonance Imaging, 10*(3), 223–232.

Tofts, P. S., Jackson, J. S., Tozer, D. J., Cercignani, M., Keir, G., MacManus, D. G., et al. (2008). Imaging cadavers: cold FLAIR and noninvasive brain thermometry using CSF diffusion. *Magn Reson Med, 59*(1), 190–195.

Tozer, D., Ramani, A., Barker, G., Davies, G., Miller, D., & Tofts, P. (2003). Quantitative magnetization transfer mapping of bound protons in multiple sclerosis. *Magnetic Resonance in Medicine, 50*(1), 83–91.

Truong, T., Chakeres, D., Scharre, D., Beversdorf, D., & Schmalbrock, P. (2006). Blipped multi gradient-echo slice excitation profile imaging (bmGESEPI) for fast T2* measurements with macroscopic B0 inhomogeneity compensation. *Magnetic Resonance in Medicine, 55*(6), 1390–1395.

Turner, R., Le Bihan, D., Maier, J., Vavrek, R., Hedges, L. K., & Pekar, J. (1990). Echo-planar imaging of intravoxel incoherent motion. *Radiology, 177*(2), 407–414.

Wang, H. Z., Riederer, S. J., & Lee, J. N. (1987). Optimizing the precision in T1 relaxation estimation using limited flip angles. *Magn Reson Med, 5*(5), 399–416.

Weigel, M., Helms, G., & Hennig, J. (2010). Investigation and modeling of magnetization transfer effects in two-dimensional multislice turbo spin echo sequences with low constant or variable flip angles at 3 T. *Magnetic Resonance in Medicine, 63*(1), 230–234.

Wild, J., Martin, W., & Allen, P. (2002). Multiple gradient echo sequence optimized for rapid, single-scan mapping of R2* at high B0. *Magnetic Resonance in Medicine, 48*(5), 867–876.

Williams, D., Detre, J., Leigh, J., & Koretsky, A. (1992). Magnetic resonance imaging of perfusion using spin inversion of arterial water. *Proceedings of the National Academy of Sciences, 89*(1), 212.

Wolff, S. D., & Balaban, R. S. (1989). Magnetization transfer contrast (MTC) and tissue water proton relaxation in vivo. *Magn Reson Med, 10*(1), 135–144.

Yarnykh, V. (2002). Pulsed Z spectroscopic imaging of cross relaxation parameters in tissues for human MRI: Theory and clinical applications. *Magnetic Resonance in Medicine, 47*(5), 929–939.

Youl, B., Hawkins, C., Morris, J., DuBoulay, E., & Tofts, P. (1992). In vivo T1 values from guinea pig brain depend on body temperature. *Magnetic resonance in medicine: official journal of the Society of Magnetic Resonance in Medicine/Society of Magnetic Resonance in Medicine, 24*(1), 170.

Young, I., Cox, I., Bryant, D., & Bydder, G. (1988). The benefits of increasing spatial resolution as a means of reducing artifacts due to field inhomogeneities. *Magnetic resonance imaging, 6*(5), 585–590.

Young, I., Hand, J., Oatridge, A., & Prior, M. (1994). Modeling and observation of temperature changes in vivo using MRI. *Magnetic resonance in medicine: official journal of the Society of Magnetic Resonance in Medicine/Society of Magnetic Resonance in Medicine, 32*(3), 358.

Yu, H., Shimakawa, A., McKenzie, C. A., Brodsky, E., Brittain, J. H., & Reeder, S. B. (2008). Multiecho water-fat separation and simultaneous R2* estimation with multifrequency fat spectrum modeling. *Magn Reson Med, 60*(5), 1122–1134.

Zerhouni, E., Parish, D., Rogers, W., Yang, A., & Shapiro, E. (1988). Human heart: tagging with MR imaging--a method for noninvasive assessment of myocardial motion. *Radiology, 169*(1), 59.

Zuo, J., Walsh, E. G., Deutsch, G., & Twieg, D. B. (2006). Rapid mapping of flow velocity using a new PARSE method. *Magn Reson Med, 55*(1), 147–152.

# F. QMRI PROCESSING

## F1. Introduction

This chapter is concerned with the qMRI processing chain, multispectral qMRI acquisitions, and methods for assessing the quality of qMRI maps.

## F2. Data Structure and Organization

### F2.1. *General Considerations: (Towards) Comprehensive qMRI*

Differentially weighted qMRI techniques rely on generating two or more sets of DA image sets that are identical in every respect including all geometric parameters --anatomic coverage, slice positions and angulations, and voxel size — and differ solely in the degree of weighting to ideally only the target qMRI parameter. Consequently, the DA images and the generated qMRI maps have the same geometric properties and differ at a fundamental level on the type of information represented: DA pixel values are unit-less parameters with qualitative meaning, while qMRI parameters bear standard physical units and represent directly a physical property of the tissue(s) of the voxel.

In general, qMRI datasets consist of n pixel values per position in the imaging grid, with the added requirement that these pixel values should be consistent with each other and therefore numerically comparable with each other. Recalling the pixel value equation, specifically:

$$
\mathrm{pv}^{\mathrm{Acq\_\lambda}}_{(i,j,sl)} = \Gamma_{(i,j,sl)} \mathrm{PD}^{(A)}_{(i,j,sl)} \mathrm{PSw}\left(\cdots \mathrm{qCV\_X}(\lambda)\cdots \middle\| \cdots \mathrm{qMRI\_par}_{(i,j,sl)}\cdots\right)
$$

$$(\text{Eq. F2.1–1})$$

a set of pixel values $\{\mathrm{pv}^{\mathrm{Acq\_1}}_{(i,j,sl)}, \ldots, \mathrm{pv}^{\mathrm{Acq\_n}}_{(i,j,sl)}\}$ is therefore suitable for qMRI processing if these have the same value of the experimental factor $\Gamma_{(i,j,sl)}$, as determined during the pulse sequence preparatory phase or pre-scanning phase.

In this context, a fully comprehensive, and therefore idealized, qMRI paradigm would generate a sufficiently large set of pixel values such that the qMRI parameters of the liquid pool, the semisolid pool, and fat can be accurately mapped. Further requirements would include: 1) high spatial resolution, 2) spatial co-registration, 3) full anatomic coverage of the targeted body part, 4) high SNR, and 5) minimal artifacts.

We express symbolically this idealized and comprehensive imaging paradigm by:

$$
\text{Object} \xrightarrow[\text{pulse sequence(s)}]{\text{qMRI}} \text{DAs}
$$

$$
= \left\{ \text{pv}_{(i,j,sl)}^{\text{Acq\_1}}, \ldots, \text{pv}_{(i,j,sl)}^{\text{Acq\_n}} \right\} \xrightarrow[\text{algorithm(s)}]{\text{qMRI}} \left\{ \begin{array}{c} \text{qMRI\_par}_{(i,j,sl)}^{(A)} \\ \oplus \\ \text{qMRI\_par}_{(i,j,sl)}^{(B)} \\ \oplus \\ \text{qMRI\_par}_{(i,j,sl)}^{(F)} \end{array} \right.
$$

$$\text{(Eq. F2.1–2)}$$

where we assume that each specific qMRI parameter --either of the liquid pool (A), the semisolid pool (B), or of fat (F) — can be determined using a subset of the acquired pixel values and the known value of the corresponding qCVs. Such an ideally comprehensive qMRI pulse sequence is yet to be described in the literature. Nevertheless, recalling the previously identified qCV matrix, *i.e.*:

$$
\text{qCV\_X} = \left\{ \begin{array}{cccc} \text{na} & \cdots & \cdots & \cdots \\ \vdots & \textbf{b-matrix} & b_{x,y,z} & \vec{V}^{(\text{enc})} \\ \vdots & (\text{TR, TI, FA}) & (\text{TE}^{(\text{SE})}, \text{TE}^{(\text{GE})}) & \text{TSL} \end{array} \right.
$$

$$\text{(Eq. F2.1–3)}$$

which contains the qCVs for probing the qMRI parameters of the liquid pool, to which we must add magnetization transfer pre-pulses for probing the semisolid pool, and spectrally selective techniques for quantifying fat, we can imagine a sufficiently fast qMRI pulse sequence that incorporates the appropriate subset of qCV variables such a qMRI-comprehensive set of DA image is generated.

At present, this is speculative and most published works have dealt with quantifying one qMRI parameter at a time *via* monospectral qMRI, whilst a few publications have dealt with multispectral qMRI techniques whereby several self-coregistered qMRI parameter maps are generated with a single pulse sequence application. Moreover, the described multispectral qMRI pulse sequences have been of limited to generating up to three primary qMRI parameters --as opposed to secondary qMRI parameters-- per scan, specifically the triads $\{PD^{(A)}_{(i,j,sl)}, T^{(A)}_{1\,(i,j,sl)}, T^{(A)}_{2(i,j,sl)}\}$ and $\{PD^{(A)}_{(i,j,sl)}, T^{(A)}_{1\,(i,j,sl)}, T^{*(A)}_{2(i,j,sl)}\}$.

Monospectral qMRI pulse sequences have a single qCV (see Eq. F4.1-3 above) the numerical value of which is varied from acquisition to acquisition. In general, these are variants of standard MRI pulse sequences and are applied sequentially in time without changing the experimental factor ($\Gamma_{(i,j,sl)}$). This can be achieved either by manually bypassing the pulse sequence preparatory phase (pre-scan) or by programming several acquisitions to run sequentially with only one pre-scanning phase at the beginning. This pre-scanning methodology accounts for most monospectral qMRI pulse sequences, the exception being multi-echo sequences for mapping $T_2$ and $T_2^*$ whereby all acquisitions occur within one TR cycle and therefore only one preparatory cycle is needed in the first place.

Multispectral qMRI pulse sequences incorporate more than one qCV in a single acquisition. For example, the mixed conventional spin-echo (mixed-CSE) pulse sequence incorporates within a single TR cycle; the principles of multi $T_1$-weighting by inversion recovery --two inversion times TI1, 2-- and of $T_2$-weighting by multi spin-echo imaging with an arbitrary number of echo times. This pulse sequence was originally described in 1987 (In den Kleef & Cuppen, 1987) by the team at Philips Medical Systems and is similar to one of the two "self-normalizing" pulse sequences described by the General Electric Medical Systems team (O'Donnell, Gore, & Adams, 1986), specifically inversion-saturation pulse sequence. For their part, the development team at Siemens Medical Systems described the "simultaneous multiple acquisition of relaxation times" (SMART) pulse sequence that used stimulated echoes (Graumann, Fischer, & Oppelt, 1986). These early multispectral qMRI pulse sequences did not incorporate fast imaging methods --such as fast spin-echo, EPI, SSFP imaging-- and were therefore clinically impractical at the time because of the long scan times. In addition, the attention of most commercial and academic research teams was concentrated on solving the basic needs of clinical MRI including many areas that are currently research active, such as body and cardiac MRI and many others.

More recently, several faster multispectral qMRI pulse sequences — described below-- with clinically acceptable scan times have been described in the literature. The inversion recovery TrueFISP (Schmitt *et al.*, 2004) pulse sequence is multispectral in PD, $T_1$, and $T_2$ and has been used for brain imaging. The mixed turbo spin-echo (mixed-TSE) is also multispectral in PD, $T_1$, and $T_2$ and has been used for abdominal applications (Farraher, Jara, Chang, Hou, & Soto, 2005; Farraher, Jara, Chang, Ozonoff, & Soto, 2006) and for brain imaging (Saito, Sakai, Ozonoff, & Jara, 2009; Suzuki, Sakai, & Jara, 2006). A third imaging method called quantification of relaxation times and proton density by twin-echo saturation-recovery turbofield echo (QRAPTEST) has been used for brain imaging (J. B. Warntjes, Dahlqvist, & Lundberg, 2007). This method was modified (J. B. M. Warntjes, Leinhard, West, & Lundberg, 2008) leading to a new method dubbed quantification of relaxation times and proton density by multiecho acquisition of a saturation-recovery using turbo spin-echo readout (QRAPMASTER). Two main issues of the original QRAPTEST method were addressed. First, a spin-echo sequence was used rather than a gradient-echo sequence to reduce susceptibility effects. Second, the maximum excitation flip angle for the QRAPTEST acquisition was typically limited to very small flip angles ($4°$–$8°$) to limit the role of the correction factor in the calculation of $T_1$. The new QRAPMASTER approach uses a multislice sequence with a long repetition time between subsequent acquisitions, thus removing the limitation of the small flip angle excitation.

### F2.2. *Multispectral qMRI Pulse Sequences*

F2.2.i. *IR-TrueFISP*

The TrueFISP pulse sequence originally described in the mid-eighties (Oppelt *et al.*, 1986) is also referred to as balanced or fully refocused SSFP, balanced FFE, or FIESTA, depending on equipment manufacturer. This pulse sequence constitutes an advantageous building block for several qMRI pulse sequences because: 1) it boasts the high scanning efficiency of the very short TR gradient-echo methods, 2) it can lead to high SNRs that are comparable to long TR sequences (K Scheffler & Lehnhardt, 2003), and 3) it has spin-echo like properties (K Scheffler & Hennig, 2003). An analysis of the phase evolution of transverse magnetization in a TrueFISP experiment showed (K Scheffler & Hennig, 2003) very close similarities to the echo formation of a spin-echo (SE) experiment. It was further shown that if dephasing between excitation pulses is below $\pm\pi$, TrueFISP exhibits a nearly complete refocusing of transverse magnetization at TE $=$ TR/2. Only signals acquired before and after TR/2 show an additional $T_2^*$ sensitivity.

By using an adiabatic inversion preparation pulse, the resulting IR-TrueFISP pulse sequence can be used for efficient $T_1$ mapping (K. Scheffler & Hennig, 2001). Furthermore, these investigators showed that useful imaging information could be obtained by acquiring signals during the transient phase before the steady state is reached. Moreover, it was shown that the signal temporal evolution during this transient period depends on both $T_1$ and $T_2$. This property was used (Schmitt, et al., 2004) for mapping the two relaxation times, which in turn enabled them to map PD with the same data.

The theory supporting PD-$T_1$-$T_2$ multispectral qMRI with this technique is reproduced in the following with minor changes in notation from the original paper (Schmitt, et al., 2004). First, it was shown that the longitudinal magnetization recovery curve is monoexponential with an apparent relaxation time given by:

$$E_1^* = (E_1 \cos^2(FA_{ex}) + E_2 \sin^2(FA_{ex}))^{-1} \qquad \text{(Eq. F2.2--1)}$$

which depends on the True-FISP flip angle and in the very short TR ($\ll T_1$, $T_2$) limit, it can be further reduced to:

$$T_1^* = \left( \frac{1}{T_1} \cos^2(FA_{ex}) + \frac{1}{T_2} \sin^2(FA_{ex}) \right)^{-1} \qquad \text{(Eq. F2.2--2)}$$

The signal just after the inversion pulse is to a good approximation given by:

$$S_0 = M_z^{(eq)} \sin\left( \frac{FA_{ex}}{2} \right) \qquad \text{(Eq. F2.2--3)}$$

and later, in the steady state is given by:

$$S_{stst} = M_z^{(eq)} \frac{(1 - E_1) \sin(FA_{ex})}{1 - (E_1 - E_2) \cos(FA_{ex}) - E_1 E_2} \qquad \text{(Eq. F2.2--4)}$$

This last equation can be further simplified in the very short TR limit, resulting in:

$$S_{stst} = M_z^{(eq)} \frac{\sin(FA_{ex})}{\left(\frac{T_1}{T_2} + 1\right) - \cos(FA_{ex})\left(\frac{T_1}{T_2} - 1\right)_1} \qquad \text{(Eq. F2.2–5)}$$

Accordingly, the recovery post IR of the magnetization during a continuous IR-TrueFISP imaging can be fitted to a three-parameter monoexponential function:

$$S(n\,TR) = 1 + \frac{S_0}{S_{stst}} = S_{stst}\left(1 - INV\,\exp\left(-\frac{nTR}{T_1^*}\right)\right) \qquad \text{(Eq. F2.2–6)}$$

The output parameters of the pixel-by-pixel fitting algorithm (qCv = nTR) are $S_{stst}$, $T_1^*$, and INV for every pixel. In this fitting equation, the inversion factor INV is a function of $T_1$ and $T_2$.

$$INV = 1 + \frac{\sin(FA_{ex}/2)}{\sin(FA_{ex})}\left(\left(\frac{T_1}{T_2} + 1\right) - \cos(FA_{ex})\left(\frac{T_1}{T_2} - 1\right)\right)$$

$$\text{(Eq. F2.2–7)}$$

In addition, from the equations above (Eq. F2.2-2, 6) we can express $T_1$ and $T_2$ in terms of the flip angle and the fitted parameters INV and $T_1^*$ and, thus leading to the following mapping equations:

$$T_{1,(i,j,sl)} = T_{1,(i,j,sl)}^*\left(\cos^2(FA_{ex}) + (A\,INV_{(i,j,sl)} + B)\sin^2(FA_{ex})\right)$$

$$\text{(Eq. F2.2–8)}$$

and,

$$T_{2,(i,j,sl)} = T_{1,(i,j,sl)}^*\left(\sin^2(FA_{ex}) + (A\,INV_{(i,j,sl)} + B)^{-1}\cos^2(FA_{ex})\right)$$

$$\text{(Eq. F2.2–9)}$$

where A and B are functions of the flip angle only, specifically:

$$A = 2(1 - \cos(FA_{ex}))^{-1}\cos(FA_{ex}/2) \qquad \text{(Eq. F2.2–10)}$$

and,

$$B = (1 + 2\cos(FA_{ex}/2) + \cos(FA_{ex}))(\cos(FA_{ex}) - 1)^{-1}$$

$$\text{(Eq. F2.2–11)}$$

Finally, Eq. F2.2-3 can be rearranged so to allow mapping of the relative proton density *via*:

$$PD_{(i,j,sl)} \propto \frac{S_{stst,(i,j,sl)}(INV_{(i,j,sl)} - 1)}{\sin(FA_{ex}/2)} \qquad (Eq. \ F2.2\text{--}12)$$

Altogether, this fitting procedure allows mapping PD, $T_1$, and $T_2$ from a single scan, which in the reported implementation generates directly acquired images of one slice at 38 different time points post inversion. The specific implementation described is as follows (Schmitt, *et al.*, 2004): after adiabatic inversion, a train of 38 image segments was acquired. Each segment was comprised of 21 phase-encoding steps, equally distributed over k-space with spacings of 12 steps. Accordingly, the recovery curve was sampled for more than 5 sec and a delay of 5 sec was introduced to allow for longitudinal relaxation before the next inversion. A square field of view of 256 mm was covered with a matrix of $252 \times 256$ pixels, yielding a total time of 2:08 min for the complete acquisition of 38 single-slice inversion recovery images. A sinc-shaped rf pulse was used, which had been optimized for a rectangular slice profile with thickness 8 mm. The same sequence was used for parameter measurements in the human brain.

### F2.2.ii. *Mixed-TSE*

As its conventional counterpart --mixed-CSE (In den Kleef & Cuppen, 1987)--, the mixed-TSE pulse sequence combines the principles of inversion recovery and multiecho sampling in one 2D multislice acquisition. For a given slice, the mixed-TSE begins with a slice-selective (adiabatic) inversion pulse. After an inversion recovery waiting time TI1 during which other slices are interrogated, a dual-echo TSE acquisition is applied. This is followed by a second waiting time TW1 after which a second dual-echo TSE acquisition is applied. Thirdly, after another waiting time TW2, the TR cycle ends. Other slices are interrogated during the waiting times TI1, TW1, and TW2 thus making it very efficient in terms of data acquisition. One mixed-TSE scan generates a set of four directly acquired images per slice. Specifically, the generated DAs are labeled as follows: IR1_E1 is $T_1$-weighted by inversion recovery and minimally T2-weighted (shortest possible $TE1_{eff}$); IR1_E2 is the T2-weighted version of IR1_E1; IR2_E1 has minimal $T_1$-weighting because TI2 is typically set very long and has minimal $T_2$-weighting; finally, IR2_E2 is the $T_2$-weighted version of IR2_E1. With these differently weighted images, maps of $T_1$ and $T_2$ can be generated with two-point qMRI algorithms, as will be explained in section F3. Furthermore, the relaxation time maps can be used to reverse the residual $T_1$- and $T_2$-weightings of one DA images, typically the IR2-E1 to generate PD maps. In this form, from one pulse sequence application, self-coregistered maps of $\{PD, T_1, \text{and } T_2\}$ can be generated. With current gradient subsystems, full coverage of the human head can be achieved with an isotropic voxel of about $1\,mm^3$ in about 12–14 minutes.

By labeling the mixed-TSE image types with the index $\alpha = IR1\_E1$, $IR1\_E2, IR2\_E1$, and $IR2\_E2$, the corresponding pixel values are:

$$pv_{(i,j,k)}^{(\alpha)} = \Omega_{(i,j,k)} \left( \frac{\pi^2 \gamma^2 \hbar^2 B_0}{k_B T} \right) PD_{(i,j,k)}^{(A)} PSw_{(i,j,k)}^{(\alpha)} \Delta V_{(i,j,k)}^{(\alpha)} \qquad \text{(Eq. F2.2–13)}$$

Where the pulse sequence weighting factors can be derived by solving the appropriate Bloch equations, leading to the following equations:

$$PSw^{IR1\_E1,2} =$$
$$= -(1 - [1 - (1 - (1 - k_{TSE})E_1(TR - TI2 - TSE_{shot}))$$
$$\times \cos[FA_{RF0}]]E_1(TI1)) \sin[FA_{RF1}]E_2(TE1, 2_{eff}) \qquad \text{(Eq. F2.2–14)}$$

and,

$$PSw^{IR2\_E1,2} = -(1 - (1 - k_{TSE})E_1(TI2 - TI1 - TSE_{shot}))E_2(TE1, 2_{eff})$$
$$\text{(Eq. F2.2–15)}$$

The standard exponential abbreviations $E_{1,2}(t) \equiv \exp(-t/T_{1,2})$ have been used.

As before, the voxel factors are expressed in terms of the voxel sensitivity functions of the appropriate readout:

$$\Delta V_{(i,j,sl)}^{Acq\_1,2} = \iiint_{\left\{\begin{array}{c}\text{Infinite}\\\text{space}\end{array}\right\}} VSF^{(Acq\_1,2)}(\vec{X}_{(i,j,sl)} - \vec{x})d^3x \qquad \text{(Eq. F2.2–16)}$$

where,

$$VSF^{(Acq\_1,2)}(\vec{X}_{(i,j,sl)} - \vec{x})$$

$$\equiv \frac{S(X_{sl} - x_{ss})}{N_p N_f} \sum_{p,f=-N/2}^{N/2} \exp \left( \frac{TE1, 2_{eff} - te_{(f,p)}^{(Acq\_1,2)}}{T_2^{(A)}(\vec{x})} \right)$$

$$\times \exp \left[ +2\pi i(\vec{X}_{(i,j)} - \vec{x}) \cdot \vec{K}_{(f,p)} \right] \qquad \text{(Eq. F2.2–17)}$$

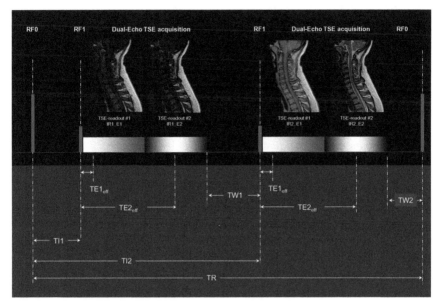

Fig. F-1. Schematic timing diagram of the mixed-TSE pulse sequence. The pulse sequence begins with the application of a slice selective inversion pulse (RF0). After an initial inversion time (TI1), a dual-echo TSE sub-sequence is applied thus generating a predominantly $T_1$-weighted image (IR1_E1) and a second image with mixed $T_1$- and $T_2$-weightings (IR1_E2). After a typically much longer second inversion time (TI2), another dual-echo-TSE sub-sequence is applied thus generating a predominantly PD-weighted image (IR2_E1) and a predominantly $T_2$-weighted image (IR2_E2). The resulting four directly acquired images are used to compute primary qMRI maps of PD, $T_1$, and $T_2$, which are self-coregistered.

The voxel-average relaxation times at each (i, j, sl)-position are computed as solutions of the root finding equations:

$$T_{1,(i,j,sl)}$$

$$= \mathbf{Root}\left\{ T_{1,(i,j,sl)} + \frac{TI1}{\ln\left[ \dfrac{1 - \mathrm{Re}\left[ \dfrac{pv^{IR1\_E1}_{(i,j,k)}}{pv^{IR2\_E1}_{(i,j,k)}} \right] (1-(1-k_{TSE})E_1(TI2-TI1-TSE_{shot}))}{(1-(1-(1-k_{TSE})E_1(TR-TI2-TSE_{shot}))(1-\kappa)\langle\cos[FA_{RF0}]\rangle_{slice})} \right]} \right\}$$

$$(\text{Eq. F2.2-18})$$

and,

$$T_{2,(i,j,sl)} = \textbf{Root} \left\{ T_{2,(i,j,sl)} + \frac{TE2_{eff} - TE1_{eff}}{\ln\left[ \left[ \frac{[pv_{(i,j,k)}^{IR2\_E2}]}{[pv_{(i,j,k)}^{IR2\_E1}]} \frac{\langle vsf_{E1}^{(PE)} \rangle_{PE}}{\langle vsf_{E2}^{(PE)} \rangle_{PE}} \right] \right]} \right\}$$

(Eq. F2.2–19)

which serve as mapping equations for this pulse sequence.

### F2.2.iii. *QRAPMASTER*

QRAPMASTER (J. B. M. Warntjes, *et al.*, 2008) is a 2D multislice saturation prepared multiecho spin-echo pulse sequence, with the capability of interrogating different slices at different times during the longitudinal recovery post saturation pulse, thus enabling $T_1$ mapping. The acquisition can be accelerated through an echo-planar imaging (EPI) technique that acquires several k-space lines per spin-echo (gradient spin-echo or GraSE). In the first pulse sequence module, a slice selective saturation pulse acts on a given slice n, followed by spoiling of the signal. In the second module, a slice-selective spin-echo acquisition of another slice m is played out; it consists of multiple echoes that are acquired to enable mapping of the transverse relaxation time. The number of echoes and the echo spacing can be adjusted freely in order to accommodate the desired dynamic range for the measurement of $T_2$. By changing the order of the slices n and m with respect to each other, the desired delay time between saturation and acquisition of a particular slice can be set. By using different delay times after a saturation pulse, the longitudinal relaxation time $T_1$ can be mapped from multiple scans. Since the number of scans and the delay times can be freely chosen, the dynamic range of $T_1$ can also be set as desired. The complete quantification measurement consists of numerous scans with different delay times TD and different echo times TE, thus enabling mapping of $T_1$ and $T_2$ independently of each other.

Key signal equations provided in this paper (J. B. M. Warntjes, *et al.*, 2008) are given below in the notation used in this book. The longitudinal magnetization at the time TD of the first multiecho module is:

$$M_z(TD) = M_z^{(eq)} - (M_z^{(eq)} - M_z(TR)\cos(FA_{sat}))\exp\left( -\frac{TD}{T_1^{(A)}} \right)$$

(Eq. F2.2–20)

and the longitudinal magnetization at the end of a TR cycle is given by:

$$M_z(TR) = M_z^{(eq)} - (M_z^{(eq)} - M_z(TD)\cos(FA_{ex}))\exp\left(-\frac{(TR-TD)}{T_1^{(A)}}\right)$$

(Eq. F2.2–21)

where $FA_{sat}$ and $FA_{ex}$ are the flip angles of the saturation and spin-echo excitation pulses respectively.

Hence, combining the two equations above, the longitudinal magnetization at TD can be expressed in terms of experimental parameters only, specifically:

$$M_z(TD)$$

$$= M_z^{(eq)}\frac{1 - (1 - \cos(FA_{sat}))\exp\left(-\frac{TD}{T_1^{(A)}}\right) - \cos(FA_{sat})\exp\left(-\frac{TR}{T_1^{(A)}}\right)}{1 - \cos(FA_{sat})\exp\left(-\frac{TR}{T_1^{(A)}}\right)\cos(FA_{ex})}$$

(Eq. F2.2–22)

This equation can be used to fit $M_z(TD)$ with respect to TD thus leading to maps of $T_1^{(A)}$ and $M_z^{(eq)}$. In addition the fitting procedure generates measures of $FA_{sat}$ for each pixel, which are denoted as $FA_{sat}^{(eff)}$. In turn, these investigators (J. B. M. Warntjes, *et al.*, 2008) use this map to generate maps of $B_1+$.

## F3. qMRI Algorithms

### F3.1. *Mapping Equations*

A qMRI algorithm is a set of computer instructions, with some steps based on solutions to the BTS equations, for generating qMRI maps using the DA images as input. In other words, given n pixel values for every pixel location, the qMRI algorithm will generate $m < n$ qMRI parameter pixel values, or symbolically:

$$\left\{pv_{(i,j,sl)}^{Acq\_1}, \ldots, pv_{(i,j,sl)}^{Acq\_n}\right\} \xrightarrow{\quad qMRI\ algorithm \quad}$$

$$\times \begin{cases} qMRI\_par\_X_{(i,j,sl)}^{(A)} & \leftrightarrow \quad \text{monospectral qMRI} \\ qMRI\_par\_X_{(i,j,sl)}^{(A)}; Y_{(i,j,sl)}^{(A)}; \ldots \text{etc.} & \leftrightarrow \quad \text{multispectral qMRI} \end{cases}$$

(Eq. F3.1–1)

where we have made a distinction between mono and multispectral qMRI.

For monospectral qMRI algorithms, all DA pixel values are used to generate one qMRI parameter pixel value:

$$\text{qMRI\_par\_X}_{(i,j,sl)}^{(A)} = \Re\left(\text{pv}_{(i,j,sl)}^{\text{Acq\_1}}, \ldots, \text{pv}_{(i,j,sl)}^{\text{Acq\_n}}, \text{qCV\_X}(1), \cdots, \text{qCV\_X}(n)\right)$$

(Eq. F3.1–2)

and for multispectral techniques, subsets of DA pixel values are used for generating different qMRI parameters at the same pixel location, accordingly:

$$\text{qMRI\_par\_X}_{(i,j,sl)}^{(A)} = \Re\left(\text{subset\_x}\left\{\text{pv}_{(i,j,sl)}^{\text{Acq\_1}}, \ldots, \text{pv}_{(i,j,sl)}^{\text{Acq\_n}}\right\},\right.$$
$$\left.\text{subset\_x}\{\text{qCV\_X}(1), \cdots, \text{qCV\_X}(n)\}\right)$$

$$\vdots$$

(Eq. F3.1–3)

$$\text{qMRI\_par\_Y}_{(i,j,sl)}^{(A)} = \Re\left(\text{subset\_y}\left\{\text{pv}_{(i,j,sl)}^{\text{Acq\_1}}, \ldots, \text{pv}_{(i,j,sl)}^{\text{Acq\_n}}\right\},\right.$$
$$\left.\text{subset\_y}\{\text{qCV\_X}(1), \ldots, \text{qCV\_X}(n)\}\right)$$

$$\vdots$$

In this case, the different qMRI maps are automatically self-coregistered and therefore useful for combined further qMRI processing, including mapping of derived qMRI parameters such as the proton density, qMRI based segmentation, and generation of images with synthetic MRI contrast, sometimes referred to as virtual MRI scanning.

### F3.2. *Model-Conforming Algorithms*

In the previous sections, we have examined several mapping equations that are specific for a given qMRI parameter and a specific qMRI pulse sequence. The corresponding mapping strategies fall into two different categories:

(1) for two point techniques where only two DA images are available, then BTS models can be derived with which the qMRI parameter is directly calculated using a mathematical formula, which may be an analytical expression or iterative such as a root finding equation.

(2) For multipoint techniques where several ($n > 2$) DA images are available, then a pixelwise multipoint fitting procedure is the logical approach. Such fitting procedure may be linear or nonlinear, and in general outputs not only the targeted qMRI parameter but also other formula parameters as well as measures of fit quality.

The general strategy for constructing qMRI algorithms, which are pixel-by-pixel algorithms, involves looping through all (i, j, sl)-positions in the imaging grid and for each pixel position, performing the following tasks on the DA pixel values:

---

***Task 1***: Read the DA pixel values for imaging grid position (i, j, sl)

***Task 2***: determine whether the DA pixel values at this position conforms to the physical BTS model.

***Task 3a***: if the DA pixel values ***do conform*** to the model, then execute the mapping equation:

$$\text{qMRI\_par\_X}^{(A)}_{(i,j,sl)} = \Re(\text{pv}^{Acq\_1}_{(i,j,sl)}, \ldots, \text{pv}^{Acq\_n}_{(i,j,sl)}, \text{qCV\_X}(1), \ldots, \text{qCV\_X}(n))$$

(Eq. F3.2–1)

***Task 3b***: if the DA pixel values ***do not conform*** to the BTS model, then compute with alternative formula or assign a pre-determined constant value.

***Task 3***: repeat from Task 1 until the last pixel of the imaging grid.

---

In a concrete example of a two point qMRI algorithm, let us reconsider the mapping formula for $T_1$ given in section F2.2.ii, specifically:

$$T_{1,(i,j,sl)} = \mathbf{Root} \left\{ T_{1,(i,j,sl)} + \frac{TI1}{\ln\left[\frac{1 - \frac{Re\left[pv^{IR1\_E1}_{(i,j,k)}\right]}{Re\left[pv^{IR2\_E1}_{(i,j,k)}\right]}(1-(1-k_{TSE})E_1(TI2-TI1-TSE_{shot}))}{(1-(1-(1-k_{TSE})E_1(TR-TI2-TSE_{shot}))(1-\kappa)\langle\cos[FA_{RF0}]\rangle_{slice})}\right]} \right\}$$

(Eq. F3.2–2)

Not all pixels necessarily conform to this model; in particular, pixels associated with tissues that are non-MR active and are therefore devoid of signal such as bone or air. For such non-conforming pixels, the formula above will give erratic answers leading to pixel dropout artifacts. Image quality is therefore significantly improved when using pixel discrimination with Task 3b. The criteria used for deciding whether a pixel does or does not conform to the physical model, are in general based on comparisons between DA pixel values among themselves as well as in relation to the noise level. For example, a given pixel may be classified as non-model conforming if its value is within the noise level, or if the differential weighting is within the noise level and therefore not sufficient to calculate the qMRI parameter.

As a second example of a multipoint algorithm, let us reconsider the case a multi-TE data set for mapping $T_2$. For this, one could use a single exponential fitting procedure using a linear least-square fitting (LLS) function whereby:

$$T_{2,(i,j,sl)} = \text{LSS} \left( \ln \left( pv_{(i,j,sl)}^{\text{Acq}\_\lambda} \right) ; TE^{\text{Acq}\_\lambda} \right) \quad \text{where} \quad \lambda \in \{1, \dots, n\}$$

$$\text{(Eq. F3.2–3)}$$

Just as for the first example above, this mapping equation can be incorporated into a model-conforming qMRI algorithm for discriminating the pixels that would give erratic results. A further question relates to deciding whether to include or not any given data point. In this particular example, data points obtained for long TEs, such that $pv < $ noise, can lead to an erroneous $T_2$ fit.

## F4. qMRI Map Quality

The quality of qMRI maps depends critically on the quality of each individual step/process of the map generation sequence including **step 1:** acquiring high-quality DA image datasets, the differential weighting of which is suitable for qMRI processing. Sufficient levels of differential weighting relative to noise are necessary for accurately estimating a qMRI parameter over a broad range of values --the dynamic range of the qMRI technique-- that ideally should encompass the biological range of tissue. **Step 2**: using a high fidelity theoretical model of the magnetization dynamics during the qMRI pulse sequence. This in turn leads to a mapping equation that is quantitatively accurate because it is true to MR physics. **Step 3**: Using a discriminating qMRI algorithm whereby pixels that do not conform to the mapping equation are appropriately processed, and last but not least, **step 4**: knowing all the actual experimental conditions of the scan, specifically the precise values of all the magnetic fields experienced by each voxel during the scan. This last requirement involves first knowing the inhomogeneities of the $B_0$ field at every point in space, as distorted by the patient, second, knowing the actual --as opposed to the nominally

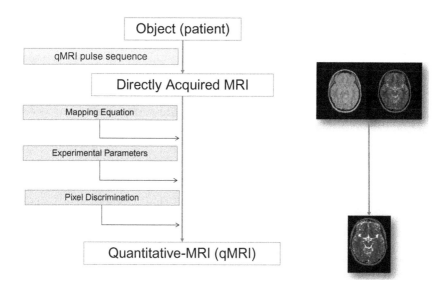

Fig. F-2.   Steps for qMRI processing.

intended-- value of each flip angle (or $B_1$) of the qMRI pulse sequence, and also knowing the actual values of the imaging gradient fields.

For several (most) qMRI parameters, **step 1** --*i.e.* acquiring high quality DA images-- is the main challenge. Such is the case for example of diffusion qMRI, which is extremely sensitive to the concomitant non-diffusional motions and is prone to several artifacts (Basser & Jones, 2002; Le Bihan, Poupon, Amadon, & Lethimonnier, 2006; Norris, 2001). Furthermore, due to gradient limitations, diffusion MRI is performed with concomitant $T_2$-weighting, hence at low SNR. To achieve the desired b-values long TEs are needed.

In a second example, acquiring artifact-free gradient-echo DA images for $T_2^*$ mapping is also challenging, in this case $B_0$ inhomogeneities are the primary culprit. In a third example, magnetization transfer qMRI is challenging because of the long scan times and high specific absorption rates (SAR) that result from applying the MT preparation pulses. In general and as a rule of thumb, the mapping of any qMRI parameter with more than two values of the appropriate qCV results in scan times that are too long for achieving the spatial resolution, anatomic coverage and high SNR expected for clinical practice. For this reason, two-point qMRI pulse sequences are particularly attractive for routine clinical use.

In other cases, **step 2** --*i.e.* developing high fidelity theoretical models of magnetization dynamics during the qMRI pulse sequence-- can be challenging as well. This is particularly true for multislice pulse sequences with many slice selective rf pulses whereby concomitant incidental magnetization transfer effects and off resonance effects lead to intricate magnetization dynamics. In such cases, the resulting mapping equations may accept only non-analytical solutions. It is pertinent at this point to recall the BTS equations (see section C3.8), which in principle should be solved for developing a truly high fidelity model of magnetization dynamics during the applications of any given qMRI pulse sequence.

$$\frac{\partial m^{(A)}}{\partial t} = -i\,2\pi\gamma B_z^{(Exp)} m^{(A)} - \frac{m^{(A)}}{T_2^{(A)}} + i\,2\pi\gamma B_{1T}^{(Exp)} M_z^{(A)}$$

$$+ \vec{\nabla}\cdot\mathbf{D}^{(A)}\cdot\vec{\nabla} m^{(A)} - \vec{v}^{(A)}\cdot\vec{\nabla} m^{(A)} \qquad \text{(Eq. F4--1)}$$

and,

$$\frac{\partial m^{(B)}}{\partial t} = -i\,2\pi\gamma B_z^{(Exp)} m^{(B)} - \frac{m^{(B)}}{T_2^{(B)}} + i\,2\pi\gamma B_{1T}^{(Exp)} M_z^{(B)} \qquad \text{(Eq. F4--2)}$$

Second, for the longitudinal magnetizations of the two pools

$$\frac{\partial M_z^{(A)}}{\partial t} = -2\pi\gamma\,\mathrm{Im}\left\{ B_{1T}^{(Exp)*} m^{(A)} \right\}$$

$$+ \left( \frac{1}{T_1^{(A)}} + R M_z^{(B\_eq)} \right) \left( M_z^{(A\_eq)} - M_z^{(A)} \right)$$

$$- R M_z^{(A\_eq)} \left( M_z^{(B\_eq)} - M_z^{(B)} \right) \qquad \text{(Eq. F4--3)}$$

and,

$$\frac{\partial M_z^{(B)}}{\partial t} = -2\pi\gamma\,\mathrm{Im}\left\{ B_{1T}^{(Exp)*} m^{(B)} \right\}$$

$$+ \left( \frac{1}{T_1^{(B)}} + R M_z^{(A\_eq)} \right) \left( M_z^{(B\_eq)} - M_z^{(B)} \right)$$

$$- R M_z^{(B\_eq)} \left( M_z^{(A\_eq)} - M_z^{(A)} \right) \qquad \text{(Eq. F4--4)}$$

Because of their intricate mathematical nature, only simplified versions of these equations are solved in each case, and these constitute imperfect models of the true magnetization dynamics.

Along the same line of reasoning, for such pulse sequences, **step 3** *i.e.* finding the precise model-conforming pixel discrimination conditions can be difficult, particularly with the increasing use of phased array coils and parallel imaging, which results in inhomogeneous noise patterns.

Lastly, **step 4** --*i.e.* knowing the actual values of the MR fields ($B_0$, $B_1$, and hence the actual flip angles) at each position of the imaging grid-- is in general time consuming as additional scans may be needed. Techniques for mapping $B_0$ and $B_1$ are reviewed later.

### F4.1. *Directly-Acquired Images: Image Quality*

The quality of a DA image dataset can be assessed by the standard quality criteria of non-quantitative MRI — (see section B3) of highest possible signal-to-noise ratio, spatial resolution, contrast resolution, and absence of artifacts--, plus the additional qMRI-specific criterion of sufficient differential-weighting-to-noise ratio. More specifically, whether there is sufficient contrast between the DAs relative to the noise levels in order to make accurate determinations of the qMRI parameter at each pixel position.

We can gain insight into the needed differential weighting by analyzing the case of a two point qMRI technique:

$$\text{CNR}_{\text{tissue}} = \frac{\left\langle \left| \text{pv}_{(i,j,sl)}^{\text{Acq\_}\lambda 1} \right| \right\rangle_{\text{ROI}} - \left\langle \left| \text{pv}_{(i,j,sl)}^{\text{Acq\_}\lambda 2} \right| \right\rangle_{\text{ROI}}}{\sigma} \qquad \text{(Eq. F4.1--1)}$$

where we assume that the noise is derived from a bivariate normal distribution with mean zero and standard deviation $\sigma$, and is the same for both acquisitions.

Using the pixel value equations,

$$\text{pv}_{(i,j,sl)}^{\text{Acq\_}\lambda 1} = \Gamma_{(i,j,sl)} \, \text{PD}_{(i,j,sl)}^{(A)} \, \text{PSw}$$
$$\times \left( \cdots \text{qCV\_X}(\lambda 1) \cdots \middle\| \cdots \text{qMRI\_par}_{(i,j,sl)} \cdots \right) \qquad \text{(Eq. F4.1--2)}$$

and,

$$\text{pv}_{(i,j,sl)}^{\text{Acq\_}\lambda 2} = \Gamma_{(i,j,sl)} \, \text{PD}_{(i,j,sl)}^{(A)} \, \text{PSw}$$
$$\times \left( \cdots \text{qCV\_X}(\lambda 2) \cdots \middle\| \cdots \text{qMRI\_par}_{(i,j,sl)} \cdots \right) \qquad \text{(Eq. F4.1--3)}$$

we find:

$$\text{CNR}_{\text{tissue}} \cong$$

$$= \Gamma_{(i,j,sl)} \, \text{PD}^{(A)}_{(i,j,sl)} \frac{\left\langle \left| \text{PSw}\left(\cdots \text{qCV\_X}(\lambda 2)\cdots \middle\| \cdots \text{qMRI\_par}_{(i,j,sl)}\cdots\right) \right| - \left| \text{PSw}\left(\cdots \text{qCV\_X}(\lambda 1)\cdots \middle\| \cdots \text{qMRI\_par}_{(i,j,sl)}\cdots\right) \right| \right\rangle_{\text{ROI}}}{\sigma}$$

$$\text{(Eq. F4.1–4)}$$

or equivalently,

$$\langle \Delta \text{PSw} \rangle_{\text{ROI}} \cong \frac{\sigma \text{CNR}_{\text{tissue}}}{\Gamma_{(i,j,sl)} \, \text{PD}^{(A)}_{(i,j,sl)}} \qquad \text{(Eq. F4.1–5)}$$

which models the needed theoretical difference in the pulse sequence weighting factors in terms of the standard deviation of the noise and CNR.

### F4.2. *Algorithm Fidelity and Nonlinearities*

Determining the uncertainty in a derived qMRI parameter as a function of the uncertainties of the pixel values --or in most cases in terms of the uncertainties of the pixel value ratios-- requires solving the BTS equations that describe the magnetization dynamics during the pulse sequence and studying the resulting mapping equation.

Different qMRI techniques have specific MR dynamics and therefore different pixel value equations and mapping equations. Hence, analyzing this problem for the general case may not be possible. We therefore analyze a two point qMRI technique to illustrate the concepts involved. Let us consider for example the $T_1$ mapping equation of the mixed-TSE pulse sequence (see Section F2.2.ii), specifically:

$$T_{1,(i,j,sl)}$$

$$= \text{Root} \left\{ T_{1,(i,j,sl)} + \frac{\text{TI1}}{\ln\left[ \frac{1 - \text{Re}(\xi_{(i,j,k)})(1 - (1 - k_{\text{TSE}}) E_1 (\text{TI2} - \text{TI1} - \text{TSE}_{\text{shot}}))}{(1 - (1 - (1 - k_{\text{TSE}}) E_1 (\text{TR} - \text{TI2} - \text{TSE}_{\text{shot}}))(1 - \kappa) \langle \cos[\text{FA}_{\text{RF0}}] \rangle_{\text{slice}}} \right]} \right\}$$

$$\text{(Eq. F4.2–1)}$$

which we have written in terms of the pixel value ratio:

$$\xi_{(i,j,sl)} \equiv \frac{\text{pv}^{\text{IR1\_E1}}_{(i,j,k)}}{\text{pv}^{\text{IR2\_E1}}_{(i,j,k)}} \qquad \text{(Eq. F4.2–2)}$$

For two-point differentially weighted qMRI techniques, pixel value ratios are
the independent experimental variable and the equations above are useful
for estimating the uncertainty in the resulting qMRI parameter *via*:

$$\Delta T_{1,(i,j,sl)} \cong \left( \frac{\partial T_{1,(i,j,sl)}}{\partial \xi_{(i,j,k)}} \right) \Delta \xi_{(i,j,k)} \qquad \text{(Eq. F4.2–3)}$$

where we have kept constant all pulse sequence parameters. Using computer
simulation for this example we can graph $T_1$ as a function of the pixel
value ratio for given values of the pulse sequence parameters, *i.e.* timing
parameters and flip angles. As shown in the figure below, $T_1$ is a monotonically decreasing nonlinear function of the pixel value ratio. Such curves are
effectively pulse sequence response functions with which pixel value ratios
are converted into the actual values of the target qMRI parameter $T_1$. In
this particular example, the range of pixel value ratios needed to cover the
full biological range --i.e. 250 ms–4,400 ms-- is $(-0.7, +0.9)$.

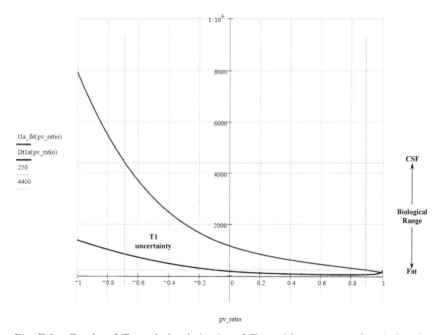

Fig. F-3.   Graphs of $T_1$ and the derivative of $T_1$ --with respect to the pixel ratio--
as functions of the pixel value ratio (see Eq. F4.2-1 and F-4.2-3). The $T_1$ uncertainty
increases for decreasing pixel value ratios, particularly for negative pixel value ratios.
Under these experimental conditions, the absolute uncertainty for a tissue with long $T_1$
is much higher than for tissues with short $T_1$s.

In summary, the analysis above provides one example for general theoretic problem of studying error propagation when processing directly acquired images to generate qMRI maps.

### F4.3. $B_0$ *Mapping Techniques*

The pixel value equation for a gradient echo pulse sequence was studied in section C5.5 for the case of ideal scanning conditions such that the phase accrued was dependent only on the $B_0$ inhomogeneities caused primarily by magnetic susceptibility differences inside the patient. In the following, we will be interested in techniques for mapping $B_0$ using a more realistic model of a scanner in which an additional phase can develop from various MR imager imperfections; for example due to eddy currents, imperfect rf penetration, and frequency-dependent transmitter and receiver nonlineariries (Kanayama, Kuhara, & Satoh, 1996). In this case, the gradient-echo pixel value equation is:

$$pv_{(i,j,sl)} = \Gamma_{(i,j,sl)} \, PD^{(A)}_{(i,j,sl)} |psw| \exp\left(-\frac{TE}{T^{(A)*}_{2,(i,j,sl)}}\right) \text{sinc}\left(2\pi\gamma\frac{\Delta B_0}{2}TE\right)$$

$$\times \exp(-i2\pi\gamma B_{0,(i,j,sl)}TE)\exp(i\varphi_{(i,j,sl)}) \qquad \text{(Eq. F4.3–1)}$$

where $\varphi$ denotes the phase secondary to scanner imperfections.

This equation suggests a straightforward general approach for $B_0$ mapping using phase differences between images acquired at different TEs. This has been accomplished using multiecho gradient-echo imaging (Kanayama, *et al.*, 1996), sequential single-echo acquisitions with shifted readout (Schneider & Glover, 1991), and with EPI pulse sequences (Roopchansingh, Cox, Jesmanowicz, Ward, & Hyde, 2003).

As a first example, if a 2D or 3D tri-echo gradient-echo pulse sequence is used, such that the time interval between the first and third echoes is equal to one chemical shift period:

$$\Delta T = \frac{\pi}{\Delta\omega_{(CS)}} = \begin{cases} 2.38\,\text{ms} & \text{at} \quad 1.5T \\ 1.19\,\text{ms} & \text{at} \quad 3.0T \end{cases} \qquad \text{(Eq. F4.3–2)}$$

This condition maintains the water and fat chemical species magnetizations independent of chemical shift phase dispersions (Schneider & Glover, 1991). Then the phase images of the first and third echoes are respectively equal to:

$$\Phi^{Echo\_1}_{(i,j,sl)} = 2\pi\gamma B_{0,(i,j,sl)}\,TE + \varphi_{(i,j,sl)} \qquad \text{(Eq. F4.3–3)}$$

and

$$\Phi_{(i,j,sl)}^{Echo\_3} = 2\pi\gamma B_{0,(i,j,sl)}(TE + \Delta T) + \varphi_{(i,j,sl)} \qquad \text{(Eq. F4.3–4)}$$

Hence, we can map $B_0$ from the phase difference:

$$B_{0,(i,j,sl)} = \frac{\Phi_{(i,j,sl)}^{Echo\_3} - \Phi_{(i,j,sl)}^{Echo\_1}}{2\pi\gamma\Delta T} \qquad \text{(Eq. F4.3–5)}$$

With this technique, the second echo is not used. Although the first and second echoes could also be used, errors may occur in the field map due to phase dispersion in the data acquisition window because these two echoes have opposing trajectories along the readout direction in k-space. Such phase dispersion is caused by microscopic inhomogeneous fields in the voxels. Furthermore, if the field inhomogeneities are large, image misregistrations artifacts can result. The misregistrations double when positive and negative readout gradient lobes are used instead of single-sided readout gradient lobes (Kanayama, *et al.*, 1996). Alternatively, $B_0$ maps can be generated using the shifted readout phase difference technique (Skinner & Glover, 1997) whereby two gradient-echo images are used with a phase difference algorithm. In this case, two single-echo acquisitions are played out sequentially with identical settings except that the second has the readout shifted by one chemical shift period. The mapping mathematics is essentially the same as described above.

The value of $\Delta T$ determines the range of phases in the phase difference map and consequently in the $B_0$ map. If the phase differences exceed the $\pm\pi$ range then, because of the phase periodicity, aliasing in $B_0$ occurs. Thus, anti-aliasing correction is necessary. In one approach (Kanayama, *et al.*, 1996), the corrected or de-aliased magnetic field map is calculated with a phase unwrapping algorithm based on the following equations:

$$\widehat{B}_{0,(i,j,sl)} = B_{0,(i,j,sl)} + \frac{n_{(i,j,sl)}}{\gamma\Delta T} \qquad \text{(Eq. F4.3–6)}$$

where n is the integer number that minimized the directional $B_0$ changes as expressed by:

$$I = \left|\frac{\partial B_{0,(i,j,sl)}}{\partial x}\right| + \left|\frac{\partial B_{0,(i,j,sl)}}{\partial y}\right| + \left|\frac{\partial B_{0,(i,j,sl)}}{\partial z}\right| \qquad \text{(Eq. F4.3–7)}$$

This is just one example of phase unwrapping algorithms and many others have been studied, *e.g.* (Song, 1995).

## F4.4. $B_1$ Mapping Techniques

In general, a transverse rf field can be decomposed into the sum of two circularly polarized and counter rotating components, termed $B_{1+}(t)\hat{e}_+$ and $B_{1-}(t)\hat{e}_-$. The former rotates in the same direction as the spins precess and the latter in the opposite direction. Magnetization excitation results from $B_{1+}(t)\hat{e}_+$, which is therefore referred to the active component of $B_1$. We therefore rewrite the on-resonance flip angle equation (see section C4.1) as:

$$\text{FA}_{(i,j,sl)} = 2\pi\gamma B_{1+}(\vec{X}_{(i,j,sl)}) \int_{-pw/2}^{pw/2} f(t)dt \qquad \text{(Eq. F4.4–1)}$$

where pw and the function f(t) are the rf pulse width and the temporal envelope respectively. This simple proportionality relationship between flip angle and $B_{1+}$ is exact for on-resonance non-slice-selective pulses and is approximate otherwise. Nevertheless, this equation is the primary link between experimental techniques that afford estimates of flip angles and $B_{1+}$ mapping.

$B_{1+}$ mapping can be accomplished with modulus based and phase sensitive differential qMRI techniques. Existing methods in the modulus based category include: 1) fitting progressively increasing flip angles (Hornak, Szumowski, & Bryant, 1988). 2) Use of stimulated echoes (Akoka, Franconi, Seguin, & Le Pape, 1993). 3) The double-angle method (DAM) (Insko & Bolinger, 1993; Stollberger & Wach, 1996) and the saturated double-angle method (SDAM) (Cunningham, Pauly, & Nayak, 2006) with which flip angle maps are generated *via* modulus pixel value ratios. 4) Signal nulling at certain flip angles (Dowell & Tofts, 2007), and 5) comparing spin-echo and stimulated echo signals (Jiru & Klose, 2006). On the other hand, existing methods of the phase sensitive category include using a series of rf pulses to produce a transverse magnetization whose phase is a function of flip angle (Morrell, 2008) and the Bloch-Siegert shift method (Sacolick, Wiesinger, Hancu, & Vogel, 2010).

Reviewing all of the above technique is beyond the scope of this book; in the following, we describe selected examples of the modulus based and of the phase sensitive categories.

The double-angle method is modulus based and uses either gradient-echo or spin-echo pulse sequences. Here we review the key equations of the gradient-echo technique, starting with the corresponding pixel value equation:

$$pv_{(i,j,sl)} =$$

$$= \Gamma_{(i,j,sl)} PD_{(i,j,sl)}^{(A)} \frac{(1 - E_1(TR))\sin(FA)}{1 - \cos(FA)E_1(TR)} \exp\left(-\frac{TE}{T_{2,(i,j,sl)}^{(A)*}}\right)$$

$$sinc\left(2\pi\gamma\frac{\Delta B_0}{2}TE\right)\exp(-i2\pi\gamma B_{0,(i,j,sl)}TE)\exp(i\varphi_{(i,j,sl)}) \quad \text{(Eq. F4.4–2)}$$

Hence, an intuitive method for $B_{1+}$ mapping is based on running a gradient-echo pulse sequence sequentially two times with the same TE and each time with a different value of the excitation flip angle. Hence, in this case, the excitation flip angle is the qCV for $B_{1+}$ and the pixel value ratios are given by:

$$\frac{pv_{(i,j,sl)}^{Acq\_1}}{pv_{(i,j,sl)}^{Acq\_2}} = \frac{\sin(FA1_{(i,j,sl)}^{(ex)})}{\sin(FA2_{(i,j,sl)}^{(ex)})}\left(\frac{1 - \cos(FA2_{(i,j,sl)}^{(ex)})E_1(TR)}{1 - \cos(FA1_{(i,j,sl)}^{(ex)})E_1(TR)}\right) \quad \text{(Eq. F4.4–3)}$$

and independent of $B_0$ inhomogeneities also. Furthermore, if the two gradient-echo pulse sequences are run in the fully unsaturated regime, *i.e.* $TR > 5\,T_{1max}$ in order to eliminate $T_1$ bias, and if the flip angles are chosen judiciously as $FA2_{(i,j,sl)}^{(ex)} = 2FA1_{(i,j,sl)}^{(ex)}$ then, a very simple flip angle mapping equation results, specifically:

$$FA1_{(i,j,sl)}^{(ex)} = \cos^{-1}\left(\frac{1}{2}\left|\frac{pv_{(i,j,sl)}^{Acq\_1}}{pv_{(i,j,sl)}^{Acq\_2}}\right|\right) \quad \text{(Eq. F4.4–4)}$$

If on the other hand, two consecutive unsaturated spin-echo pulse sequences are acquired such that the first uses $(FA1_{(i,j,sl)}^{(ex)} - 2FA1_{(i,j,sl)}^{(ex)})$ and the second with twice the flip angles $(2FA1_{(i,j,sl)}^{(ex)} - 4FA1_{(i,j,sl)}^{(ex)})$, then the corresponding flip angle mapping equation is:

$$FA1_{(i,j,sl)}^{(ex)} = \sec^{-1}\left(\sqrt[3]{9\frac{\left|pv_{(i,j,sl)}^{Acq\_1}\right|}{\left|pv_{(i,j,sl)}^{Acq\_2}\right|}}\right) \quad \text{(Eq. F4.4–5)}$$

In summary, double-angle $B_{1+}$ mapping techniques are theoretically insensitive to $B_0$ inhomogeneities and the mapping equations are computationally simple. The experimental results of a recent investigation

(Wang, Mao, Qiu, Smith, & Constable, 2006) demonstrate that rf pulse shapes significantly affect the flip angle distribution and calibration factors. Furthermore, off-resonance rf excitations, $B_0$ nonuniformities, and slice-selection gradients can lead to degradations in the signal intensities of the images used to map the flip angle, and potentially introduce a bias and increased variance in the measured flip angles. The main drawback with these two double-angle techniques is the need for operating in the long TR regime, which results in long scan times. An alternative technique designed to operate in the saturated regime and therefore termed saturated double-angle method (SDAM) has been described (Cunningham, *et al.*, 2006).

We now succinctly review a recently developed (Sacolick, *et al.*, 2010) phase sensitive $B_{1+}$ mapping technique that uses either gradient-echo or spin-echo pulse sequences with a modification consisting in applying off-resonance rf pulses after excitation. In the case of a gradient-echo pulse sequence, one such off-resonance pulse is applied at frequency $\omega_{rf}$ relative to the master frequency $\omega_0$ and two pulses at $\pm\omega_{rf}$ are applied for spin-echo sequences; one before and one after the refocusing rf pulse. Applying this off-resonance radiation causes a change in the precession frequency of the transverse magnetization while the pulse(s) are applied. Hence, the net cumulative effect is a phase shift in the measured MR signals and pixel values, termed the Bloch-Siegert phase shift (Bloch & Siegert, 1940). This effect can be explained as an additional contribution to the static $B_0$ field that arises from the off-resonance component of the rf field (Sacolick, *et al.*, 2010). In the ideal case of a perfectly homogeneous $B_0$, the Bloch-Siegert phase shift is given by:

$$\Phi_{BS} = \int_0^T \frac{(2\pi\gamma B_1(t))^2}{2\omega_{rf}(t)} dt \qquad \text{(Eq. F4.4–6)}$$

where $B_1(t)$ and $\omega_{rf}$ are the amplitude and frequency --which can be made time dependent-- in the rotating frame of the applied off-resonance radiation. Analysis of this effect in the more realistic case that includes $B_0$ nonuniformities, the Bloch-Siegert phase shift is given by the following approximate equation (Sacolick, *et al.*, 2010):

$$\Phi_{BS} \approx \int_0^T \frac{(2\pi\gamma B_1(t))^2}{2\omega_{rf}(t)} dt - \int_0^T \frac{(2\pi\gamma B_1(t))^2 \omega_{B0}}{2\omega_{rf}^2(t)} dt + O(\omega_{B0}^2)$$

$$\text{(Eq. F4.4–7)}$$

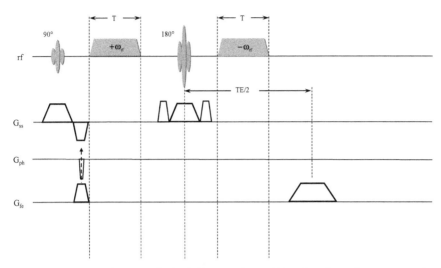

Fig. F-4. Timing diagram of Bloch-Siegert phase shift spin-echo pulse sequence for mapping $B_1$.

As all qMRI techniques by differential weighting, this phase-based method requires taking the difference of two scans to remove additional, undesired phase effects in the image. Thus these investigators (Sacolick, *et al.*, 2010) used two scans, one with the rf pulse at $+\omega_{rf}$, and one at $-\omega_{rf}$, thus enabling them to map $B_1$ with the equation above. The difference in phase between such two scans gives the pure Bloch-Siegert phase shift, whereby concomitant phases due to transmit excitation and receive imperfections, other sequence-related phases, and the phase shift from off-resonance $B_0$, are removed. All these phase factors are the same in both scans.

## References

Akoka, S., Franconi, F., Seguin, F., & Le Pape, A. (1993). Radiofrequency map of an NMR coil by imaging. *Magnetic resonance imaging, 11*(3), 437–441.

Basser, P. J., & Jones, D. K. (2002). Diffusion tensor MRI: theory, experimental design and data analysis–a technical review. *NMR in Biomedicine, 15*(7 8), 456–467.

Bloch, F., & Siegert, A. (1940). Magnetic resonance for nonrotating fields. *Physical Review, 57*, 522–527.

Cunningham, C., Pauly, J., & Nayak, K. (2006). Saturated double-angle method for rapid B1+ mapping. *Magnetic resonance in medicine: official journal of the Society of Magnetic Resonance in Medicine/Society of Magnetic Resonance in Medicine, 55*(6), 1326.

Dowell, N., & Tofts, P. (2007). Fast, accurate, and precise mapping of the RF field in vivo using the 180° signal null. *Magnetic Resonance in Medicine, 58*(3), 622–630.

Farraher, S. W., Jara, H., Chang, K. J., Hou, A., & Soto, J. A. (2005). Liver and spleen volumetry with quantitative MR imaging and dual-space clustering segmentation. *Radiology, 237*(1), 322–328.

Farraher, S. W., Jara, H., Chang, K. J., Ozonoff, A., & Soto, J. A. (2006). Differentiation of hepatocellular carcinoma and hepatic metastasis from cysts and hemangiomas with calculated T2 relaxation times and the T1/T2 relaxation times ratio. *J Magn Reson Imaging, 24*(6), 1333–1341.

Graumann, R., Fischer, H., & Oppelt, A. (1986). A new pulse sequence for determining T1 and T2 simultaneously. *Med Phys, 13*(5), 644–647.

Hornak, J. P., Szumowski, J., & Bryant, R. G. (1988). Magnetic field mapping. *Magnetic Resonance in Medicine, 6*(2), 158–163.

In den Kleef, J. J., & Cuppen, J. J. (1987). RLSQ: T1, T2, and rho calculations, combining ratios and least squares. *Magn Reson Med, 5*(6), 513–524.

Insko, E., & Bolinger, L. (1993). Mapping of the radiofrequency field. *Journal of magnetic resonance. Series A, 103*(1), 82–85.

Jiru, F., & Klose, U. (2006). Fast 3D radiofrequency field mapping using echo-planar imaging. *Magnetic Resonance in Medicine, 56*(6), 1375–1379.

Kanayama, S., Kuhara, S., & Satoh, K. (1996). In vivo rapid magnetic field measurement and shimming using single scan differential phase mapping. *Magnetic Resonance in Medicine, 36*(4), 637–642.

Le Bihan, D., Poupon, C., Amadon, A., & Lethimonnier, F. (2006). Artifacts and pitfalls in diffusion MRI. *Journal of Magnetic Resonance Imaging, 24*(3), 478–488.

Morrell, G. (2008). A phase-sensitive method of flip angle mapping. *Magnetic Resonance in Medicine, 60*(4), 889–894.

Norris, D. (2001). Implications of bulk motion for diffusion-weighted imaging experiments: effects, mechanisms, and solutions. *Journal of Magnetic Resonance Imaging, 13*(4), 486–495.

O'Donnell, M., Gore, J. C., & Adams, W. J. (1986). Toward an automated analysis system for nuclear magnetic resonance imaging. I. Efficient pulse sequences for simultaneous T1-T2 imaging. *Med Phys, 13*(2), 182–190.

Oppelt, A., Graumann, R., Barfuss, H., Fischer, H., Hartl, W., & Schajor, W. (1986). FISP: a new fast MRI sequence. *Electromedica, 54*(1), 15–18.

Roopchansingh, V., Cox, R. W., Jesmanowicz, A., Ward, B. D., & Hyde, J. S. (2003). Single-shot magnetic field mapping embedded in echo-planar time-course imaging. *Magnetic Resonance in Medicine, 50*(4), 839–843.

Sacolick, L., Wiesinger, F., Hancu, I., & Vogel, M. (2010). B1 mapping by Bloch-Siegert shift. *Magnetic Resonance in Medicine, 63*(5), 1315–1322.

Saito, N., Sakai, O., Ozonoff, A., & Jara, H. (2009). Relaxo-volumetric multispectral quantitative magnetic resonance imaging of the brain over the human lifespan: global and regional aging patterns. *Magn Reson Imaging, 27*(7), 895–906.

Scheffler, K., & Hennig, J. (2001). T(1) quantification with inversion recovery TrueFISP. *Magn Reson Med, 45*(4), 720–723.

Scheffler, K., & Hennig, J. (2003). Is TrueFISP a gradient-echo or a spin-echo sequence? *Magnetic Resonance in Medicine, 49*(2), 395–397.

Scheffler, K., & Lehnhardt, S. (2003). Principles and applications of balanced SSFP techniques. *European radiology, 13*(11), 2409–2418.

Schmitt, P., Griswold, M. A., Jakob, P. M., Kotas, M., Gulani, V., Flentje, M., et al. (2004). Inversion recovery TrueFISP: quantification of T(1), T(2), and spin density. *Magn Reson Med, 51*(4), 661–667.

Schneider, E., & Glover, G. (1991). Rapid in vivo proton shimming. *Magnetic resonance in medicine: official journal of the Society of Magnetic Resonance in Medicine/Society of Magnetic Resonance in Medicine, 18*(2), 335.

Skinner, T. E., & Glover, G. H. (1997). An extended two-point Dixon algorithm for calculating separate water, fat, and B0 images. *Magn Reson Med, 37*(4), 628–630.

Song, M. (1995). Phase unwrapping of MR phase images using Poisson equation. *IEEE Transactions on Image Processing, 4*(5), 667–676.

Stollberger, R., & Wach, P. (1996). Imaging of the active B1 field in vivo. *Magnetic resonance in medicine: official journal of the Society of Magnetic Resonance in Medicine/Society of Magnetic Resonance in Medicine, 35*(2), 246.

Suzuki, S., Sakai, O., & Jara, H. (2006). Combined volumetric T1, T2 and secular-T2 quantitative MRI of the brain: age-related global changes (preliminary results). *Magn Reson Imaging, 24*(7), 877–887.

Wang, J., Mao, W., Qiu, M., Smith, M., & Constable, R. (2006). Factors influencing flip angle mapping in MRI: RF pulse shape, slice-select gradients, off-resonance excitation, and B0 inhomogeneities. *Magnetic Resonance in Medicine, 56*(2), 463–468.

Warntjes, J. B., Dahlqvist, O., & Lundberg, P. (2007). Novel method for rapid, simultaneous T1, T*2, and proton density quantification. *Magn Reson Med, 57*(3), 528–537.

Warntjes, J. B. M., Leinhard, O. D., West, J., & Lundberg, P. (2008). Rapid magnetic resonance quantification on the brain: Optimization for clinical usage. *Magnetic Resonance in Medicine, 60*(2), 320–329.

# G. INTRODUCTION TO APPLICATIONS OF QMRI

## G1. Introduction

This chapter is concerned with several clinical and scientific applications that are possible when qMRI data are available. At a basic level, a qMRI dataset is a quantitative digital model of a body part of the patient and therefore it can be used as a virtual patient for computer simulation. Of particular interest is simulating the process of MRI scanning itself with a method known as synthetic MRI, the process of surgical excision *via* tissue/organ segmentation, and computer-aided diagnosis *via* volumetry and qMRI spectral analyses.

## G2. Image Synthesis: Virtual MRI Scanning

Synthetic MRI was originally described in the early 1980s (Ortendahl, Hylton, Kaufman, & Crooks, 1984; S. J. Riederer *et al.*, 1984). It was the focus of several studies during the mid eighties (Bobman, Riederer, Lee, Suddarth, Wang, Drayer, *et al.*, 1985; Bobman, Riederer, Lee, Suddarth, Wang, & MacFall, 1985; Bobman *et al.*, 1986; Lee & Riederer, 1986a, 1986b, 1987; Lee, Riederer, Bobman, Farzaneh, & Wang, 1986; Lee, Riederer, Bobman, Johnson, & Farzaneh, 1986; MacFall, Riederer, & Wang, 1986; S. Riederer, Bobman, Lee, Farzaneh, & Wang, 1986; S. J. Riederer, Lee, Farzaneh, Wang, & Wright, 1986). More recently, with the advent of faster pulse sequences, interest in this qMRI application is gradually reemerging as evidenced from recent articles (Andreisek *et al.*, 2010; Deoni, Rutt, & Peters, 2006) and patents (Busse, Pauly, & Zaharchuk, 2007; Jara, 2004, 2006; Kabasawa, 2004).

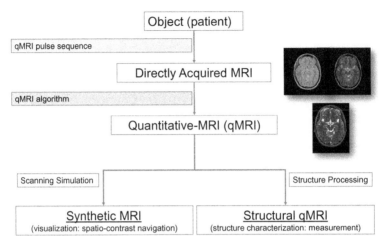

Fig. G-1.   Applications of qMRI.

In concrete terms, synthetic MRI refers to the process of generating images with arbitrary synthesized weightings to a given qMRI parameter. At a conceptual level, synthetic MRI can be understood as a computerized process that simulates MRI scanning as applied to a virtual patient that is represented by qMRI datasets. A minimum of two maps per slice are needed in order to generate a series of synthetic images; one of the proton density that serves as base layer and one of the actual qMRI parameter, the weighting of which we intend to synthesize. For example, in order to generate a series of synthetic $T_1$-weighted images, a PD and a $T_1$ map are needed.

Synthetic MRI can be understood by examining the pixel value equation of section E2.1, which for image synthesis purposes can be used in the simpler form:

$$pv_{(i,j,sl)}^{Synth} = \Gamma_{(i,j,sl)} PD_{(i,j,sl)}^{(A)} PSw(\ldots qCV\_X \ldots || \ldots qMRI\_par_{(i,j,sl)} \ldots)$$
$$\text{(Eq. F4.4–1)}$$

Accordingly, using the appropriate qMRI maps as input, weighted images by any pulse sequence can be generated with a suitable pulse sequence weighting factor that is true to pulse sequence's MR dynamics. The pixel value equations for four of the most used pulse sequences were presented in section C6.5, and are reproduced here adapted for pixel-by-pixel image synthesis by indicating the pixel indices where appropriate.

First, for the SSFP pulse sequence:

$$\text{pv}_{(i,j,sl)} = \Gamma_{(i,j,sl)} \text{PD}^{(A)}_{(i,j,sl)}$$

$$\times \frac{(1 - E^{(A)}_{1,(i,j,sl)}(\text{TR})) \sin(\text{FA})}{(1 - E^{(A)}_{1,(i,j,sl)}(\text{TR}) E^{(A)}_{2,(i,j,sl)}(\text{TR}) - (E^{(A)}_{1,(i,j,sl)}(\text{TR}) - E^{(A)}_{2,(i,j,sl)}(\text{TR})) \cos(\text{FA})}$$

$$\times \exp\left(-\frac{\text{TE}}{T^{(A)*}_{2,(i,j,sl)}}\right) \qquad \text{(Eq. F4.4-2)}$$

second, for the spoiled gradient-echo:

$$\text{pv}_{(i,j,sl)} = \Gamma_{(i,j,sl)} \text{PD}^{(A)}_{(i,j,sl)}$$

$$\times \frac{(1 - E^{(A)}_{1,(i,j,sl)}(\text{TR})) \sin(\text{FA})}{(1 - E^{(A)}_{1,(i,j,sl)}(\text{TR}) \cos(\text{FA})} \exp\left(-\frac{\text{TE}}{T^{(A)*}_{2,(i,j,sl)}}\right)$$

$$\text{(Eq. F4.4-3)}$$

third, for diffusion-weighted spin-echo:

$$\text{pv}_{(i,j,sl)} = \Gamma_{(i,j,sl)} \text{PD}^{(A)}_{(i,j,sl)} \left(1 - 2\exp\left(-\frac{(\text{TR} - \text{TE}/2)}{T^{(A)}_{1,(i,j,sl)}}\right)\right.$$

$$\left. + \exp\left(-\frac{\text{TR}}{T^{(A)}_{1,(i,j,sl)}}\right)\right) \exp\left(-\frac{\text{TE}}{T^{(A)}_{2,(i,j,sl)}}\right) \exp(-b D_{(i,j,sl)})$$

$$\text{(Eq. F4.4-4)}$$

and fourth, for inversion recovery spin-echo:

$$\text{pv}_{(i,j,sl)} = \Gamma_{(i,j,sl)} \text{PD}^{(A)}_{(i,j,sl)} \left(1 - 2\exp\left(-\frac{\text{TI}}{T^{(A)}_{1,(i,j,sl)}}\right)\right.$$

$$\left. + 2\exp\left(-\frac{(\text{TR} - \text{TE}/2)}{T^{(A)}_{1,(i,j,sl)}}\right) - \exp\left(-\frac{\text{TR}}{T^{(A)}_{1,(i,j,sl)}}\right)\right)$$

$$\times \exp\left(-\frac{\text{TE}}{T^{(A)}_{2,(i,j,sl)}}\right) \qquad \text{(Eq. F4.4-5)}$$

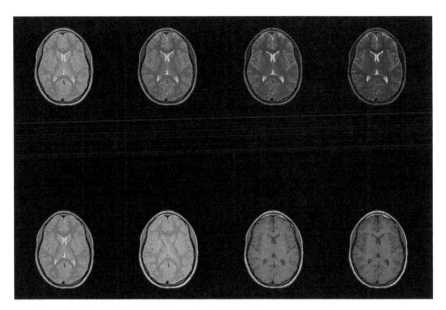

Fig. G-2 Synthetic MRI with the mixed-TSE pulse sequence. Two contrast series are shown: first, a $T_2$-weighted series (top row) with simulated echo times of $TE = 0$, 40, 80, and 120ms. Second, a $T_1$-weighted series (bottom row) with simulated repetition times of $TR = 40$, 2, 1, and 0.7s. These are selected examples only: TR and TE can be adjusted at will thus allowing for continuous change of contrast weightings.

We note in passing that synthetic MRI allows for continuously varying the level of contrast weighting and this offers a capability for visualizing subtle structures in an image by fine-tuning the contrast between such image features relative to its surroundings.

## G3. Characterization of Organs and Tissue Types

### G3.1. *qMRI Space*

Voxels can be ordered either by the spatial coordinates (i, j, sl) or by the qMRI pixel values themselves, which being well defined scientific quantities can therefore be used as generalized coordinates. Accordingly, we can define an additional space for the imaging data termed hereafter qMRI space, the dimensionality of which equals number of qMRI parameters available. For example, if a dataset is multispectral in PD, $T_1$, and $T_2$, a three-dimensional qMRI space results. Furthermore, if such three-fold multispectral dataset is coregistered to an additional qMRI dataset --*e.g.* diffusion coefficient

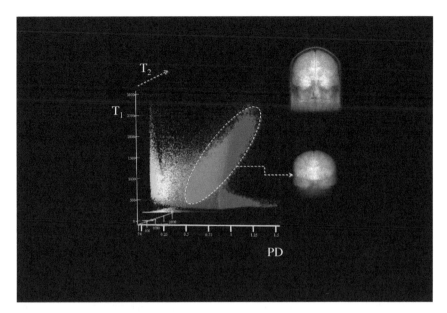

Fig. G-3. Three-dimensional qMRI space of a human head in which the coordinates of each point are the relative proton density (PD), $T_1$, and $T_2$. The top insert is a coronal projection of the whole head and the bottom insert is a coronal projection of the all-intracranial tissues and corresponds to the segment of qMRI space in the ellipsoid, as segmented with a dual-clustering algorithm.

maps--, then the qMRI space becomes four-dimensional and so forth. In short, the dimension of qMRI space is the spectrality of the qMRI data, and therefore can be much larger than that of geometric space.

## G3.2. *Segmentation*

The information contained in either space --*i.e.* geometric space and qMRI space-- are independent of the other and they are therefore complementary. Such information complementarity can be useful for uniquely identifying voxels, for discriminating between voxels, and therefore for developing highly automated segmentation algorithms. The operational principles of such an algorithm termed dual-clustering (Farraher, Jara, Chang, Hou, & Soto, 2005; Watanabe, Sakai, Norbash, & Jara, 2010) is to interrogate pixel-by-pixel as to whether it is contained in a certain qMRI subvolume and also whether it is surrounded or clustered in geometric space by pixels that are qMRI-similar. Accordingly,

$$\text{pixel}_{(i,j,sl)} \in \left\{ \begin{array}{c} \text{tissue} \\ \text{Segment} \end{array} \right\} \quad \text{if}$$

$$\times \left| \begin{array}{c} \text{qMRI\_param}^{(1)}_{(i,j,sl)} \in \text{range}\{\text{qMRI\_param}^{(1)}\} \\ \text{and} \\ \downarrow \\ \vdots \\ \text{qMRI\_param}^{(N)}_{(i,j,sl)} \in \text{range}\{\text{qMRI\_param}^{(N)}\} \\ \downarrow \\ \text{pixel} \in \text{cluster}\{\text{qMRI-similar pixel}\} \end{array} \right. \qquad \text{(Eq. G3.2–1)}$$

The first N conditional Boolean statements of this dual-space clustering algorithm are used to prescribe an initial volume of acceptance in qMRI space. Furthermore, the geometric connectivity condition in image space –*i.e.* the last statement in the equation above—further interrogates the surrounding pixels of a cluster for qMRI-similarity. As defined above, two pixels are qMRI-similar if they are both contained in the qMRI subvolume defined in the first N Boolean statements; more specifically, the last statement of this equation is:

$$\text{pixel}_{(i,j,sl)} \in \left\{ \begin{array}{c} \text{tissue} \\ \text{Segment} \end{array} \right\} \quad \text{if}$$

$$\times \sum_{a=j-\alpha}^{j+\alpha} \sum_{b=k-\alpha}^{k+\alpha} (\text{similarity\_index})_{a,b} \geq \text{CN}$$

$$\text{(Eq. G3.2–2)}$$

This last statement includes two segmentation control parameters to be prescribed *a priori* by the operator, specifically the spatial cluster area $(2\alpha + 1)^2$ expressed in units of pixels and the similarly clustering condition percentage (CN) required for accepting a given voxel to the desired segment. This type of algorithms is very effective for segmenting well circumscribed organs such as the liver, the spleen, and the brain in its totality (Watanabe, *et al.*, 2010) with either one or three qMRI channels.

## G3.3. *Volumetry and qMRI Spectroscopy*

The output of a segmentation algorithm can be saved as a binary mask whereby pixels are assigned a value of one if they belong to the segment and zero otherwise. The volume of a segment is then easily calculated as the product of the voxel volume and the number of pixels in the segment:

$$V = \Delta V \sum_{i,j,sl}^{Nx,Ny,Ns} \text{mask}_{(i,j,sl)} \qquad \text{(Eq. G3.3--1)}$$

The qMRI parameter segments can be calculated by multiplying the qMRI maps with the mask on a pixel-by-pixel basis.

$$\text{qMRI\_param}_{(i,j,sl)}^{(\text{Seg})} = \text{mask}_{(i,j,sl)}^{(\text{Seg})} \text{qMRI\_param}_{(i,j,sl)} \qquad \text{(Eq. G3.3--2)}$$

The segment mask can in turn be used for generating the segmental frequency histograms, specifically one per qMRI space dimension. This can be accomplished by first multiplying the three-dimensional mask matrix with the qMRI parameter matrix on a pixel-by-pixel basis for generating qMRI maps of the segment only. Then one needs to generate a one-dimensional array containing only the non-null elements and arrange these in increasing order.

$$\text{qMRI\_param}_{(i,j,sl)}^{(\text{Seg})} \longrightarrow \text{qMRI\_vector}^{(\text{Seg})} \qquad \text{(Eq. G3.3--3)}$$

The spatial information is lost with this vectorization operation, and the generated data vector bears only one index running from one to the total number of pixels in the segment. A histogram can then be generated from this non-null one-dimensional array by counting the number of pixels per parameter unit interval or bin.

$$h^{(\text{Seg})} = \text{histo}(\text{bins}, \text{qMRI\_vector}^{(\text{Seg})}) \qquad \text{(Eq. G3.3--4)}$$

where the first argument (bins) of the histogram function is a vector containing the positions of each bin along the axis of qMRI space. The bin width should be chosen judiciously or noise could be added in order to avoid artifactual spikes that can arise when dividing of two series of integers (Tozer & Tofts, 2003). As mentioned above, the spatial information is lost when constructing a histogram and the qMRI pixel values are now labeled according to their position along the qMRI axis for that parameter. Hence a histogram is a measure of likelihood for pixels of a segment to be within a certain interval of a given qMRI parameter.

Empirically, segments containing a single normal tissue type tend to have symmetric histograms with nearly gaussian shapes and the histograms of multi-tissue/multi-organ segments tend to be multimodal with each mode representing in most cases a single tissue type. Accordingly, we can think of a histogram as the qMRI spectrum of a body part, which can be processed

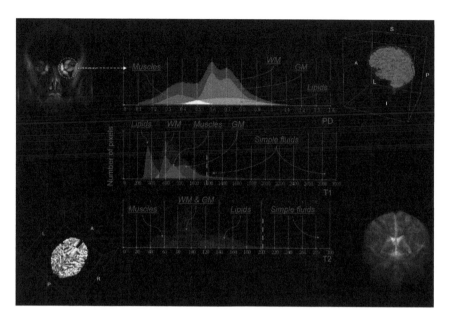

Fig. G-4.　Histogram representation of the qMRI data.

or deconvolved into primary tissue types *via* spectroscopic methods. For each qMRI parameter, the shapes and positions of such "base histograms" or primary spectral features depend on the age and the state of health of the tissue as well as on experimental conditions; primarily the field strength $B_0$, the temperature of the patient, and the quality of the qMRI technique. As discussed in Chapter E, several techniques are available for quantifying each qMRI parameter and achieving data consistency among the techniques is still works in progress.

## References

Andreisek, G., White, L. M., Theodoropoulos, J. S., Naraghi, A., Young, N., Zhao, C. Y., *et al.* (2010). Synthetic-echo time postprocessing technique for generating images with variable T2-weighted contrast: diagnosis of meniscal and cartilage abnormalities of the knee. *Radiology, 254*(1), 188–199.

Bobman, S. A., Riederer, S. J., Lee, J. N., Suddarth, S. A., Wang, H. Z., Drayer, B. P., *et al.* (1985). Cerebral magnetic resonance image synthesis. *AJNR Am J Neuroradiol, 6*(2), 265–269.

Bobman, S. A., Riederer, S. J., Lee, J. N., Suddarth, S. A., Wang, H. Z., & MacFall, J. R. (1985). Synthesized MR images: comparison with acquired images. *Radiology, 155*(3), 731–738.

Bobman, S. A., Riederer, S. J., Lee, J. N., Tasciyan, T., Farzaneh, F., & Wang, H. Z. (1986). Pulse sequence extrapolation with MR image synthesis. *Radiology, 159*(1), 253–258.

Busse, R., Pauly, J., & Zaharchuk, G. (2007). Method for generating T1-weighted magnetic resonance images and quantitative T1 maps. US Patent 7,276,904.

Deoni, S. C., Rutt, B. K., & Peters, T. M. (2006). Synthetic T1-weighted brain image generation with incorporated coil intensity correction using DESPOT1. *Magn Reson Imaging, 24*(9), 1241–1248.

Farraher, S. W., Jara, H., Chang, K. J., Hou, A., & Soto, J. A. (2005). Liver and spleen volumetry with quantitative MR imaging and dual-space clustering segmentation. *Radiology, 237*(1), 322–328.

Jara, H. (2004). Synthetic images for a magnetic resonance imaging scanner using linear combination of source images to generate contrast and spatial navigation. United States patent number US 6,823,205 B1.

Jara, H. (2006). Synthetic images for a magnetic resonance imaging scanner using linear combinations of source images. United States patent number US 7,002,345.

Kabasawa, H. (2004). Magnetic resonance imaging apparatus and magnetic resonance imaging method. US 2005/0134266 A1.

Lee, J. N., & Riederer, S. J. (1986a). A modified saturation-recovery approximation for multiple spin-echo pulse sequences. *Magn Reson Med, 3*(1), 132–134.

Lee, J. N., & Riederer, S. J. (1986b). Optimum acquisition times of two spin echoes for MR image synthesis. *Magn Reson Med, 3*(4), 634–638.

Lee, J. N., & Riederer, S. J. (1987). The contrast-to-noise in relaxation time, synthetic, and weighted-sum MR images. *Magn Reson Med, 5*(1), 13–22.

Lee, J. N., Riederer, S. J., Bobman, S. A., Farzaneh, F., & Wang, H. Z. (1986). Instrumentation for rapid MR image synthesis. *Magn Reson Med, 3*(1), 33–43.

Lee, J. N., Riederer, S. J., Bobman, S. A., Johnson, J. P., & Farzaneh, F. (1986). The precision of TR extrapolation in magnetic resonance image synthesis. *Med Phys, 13*(2), 170–176.

MacFall, J. R., Riederer, S. J., & Wang, H. Z. (1986). An analysis of noise propagation in computed T2, pseudodensity, and synthetic spin-echo images. *Med Phys, 13*(3), 285–292.

Ortendahl, D. A., Hylton, N. M., Kaufman, L., & Crooks, L. E. (1984). Signal to noise in derived NMR images. *Magn Reson Med, 1*(3), 316–338.

Riederer, S., Bobman, S., Lee, J., Farzaneh, F., & Wang, H. (1986). Improved precision in calculated T1 MR images using multiple spin-echo acquisition. *Journal of computer assisted tomography, 10*(1), 103.

Riederer, S. J., Lee, J. N., Farzaneh, F., Wang, H. Z., & Wright, R. C. (1986). Magnetic resonance image synthesis. Clinical implementation. *Acta Radiol Suppl, 369*, 466–468.

Riederer, S. J., Suddarth, S. A., Bobman, S. A., Lee, J. N., Wang, H. Z., & MacFall, J. R. (1984). Automated MR image synthesis: feasibility studies. *Radiology, 153*(1), 203–206.

Tozer, D. J., & Tofts, P. S. (2003). Removing spikes caused by quantization noise from high-resolution histograms. *Magn Reson Med, 50*(3), 649–653.

Watanabe, M., Sakai, O., Norbash, A. M., & Jara, H. (2010). Accurate brain volumetry with diffusion-weighted spin-echo single-shot echo-planar-imaging and dual-clustering segmentation: comparison with volumetry-validated quantitative magnetic resonance imaging. *Med Phys, 37*(3), 1183–1190.

CPSIA information can be obtained
at www.ICGtesting.com
Printed in the USA
LVHW060017190319
611096LV00002B/16/P